Learning C with Fractals

ROGER T. STEVENS

ACADEMIC PRESS, INC.
Harcourt Brace Jovanovich, Publishers

Boston San Diego New York
London Sydney Tokyo Toronto

This book is printed on acid-free paper. ∞

ACADEMIC PRESS, INC.
1250 Sixth Avenue, San Diego, CA 92101-4311

United Kingdom Edition published by
ACADEMIC PRESS LIMITED
24–28 Oval Road, London NW1 7DX

Library of Congress Cataloging-in-Publication Data

Stevens, Roger T., date-
Learning C with fractals / Roger T. Stevens.
p. cm.
ISBN 0-12-668315-8 (alk. paper)
1. C (Computer program language) 2. Fractals—Data processing.
I. Title.
QA76.73.C15S737 1993
005.13'3—dc20 92-35032
CIP

Printed in the United States of America
93 94 95 96 MM 9 8 7 6 5 4 3 2 1

Acknowledgments

All of the Software in this book was written in Borland C++, furnished by Borland International, 4385 Scotts Valley Drive, Scotts Valley, California 95066.

Computer graphics were tested on a Powergraph ERGO-VGA board with 1 megabyte of memory furnished by STB Systems, Inc., 1651 N. Glenville, Suite 210, Richardson, Texas 75081.

Table of Contents

List of Color Plates

1

Introduction

When my book *Fractal Programming in C* was published, I received a number of calls that all started out about the same way. "I always wanted to learn how to program in C," the caller would say, "but I never had any inducement until I saw your book. I bought it and tried to learn C, but none of your programs work." In every case, as I further questioned the caller, I was able to straighten out his problems, which were caused not by anything wrong with the programs but due to the fact that he didn't know how to program in the C language. From these calls, I realized that a book was needed that would teach the C language and that would use example programs that were graphic and interesting to capture the attention of those people who always wanted to program in C but found the example programs in conventional C texts so dull and boring that they could never sustain interest in learning how to use the language properly. If you're one of these people, this book will teach you the C language and at the same time introduce you to a number of fractals that you can generate on your own display.

Hardware Requirements

The example programs in this book are designed to be run on an IBM PC or clone having a VGA card and monitor. You really need a good color display to show the detail and beauty of the fractals that are presented. If you have an EGA card and monitor, you can modify the programs so that they will work, but I'd recommend that you switch to the VGA, because in the near future it will be a minimum requirement for many programs. You don't absolutely need a hard disk to run these programs, but you will find that most of the C compilers currently available are so large that they require a hard disk. Many of the fractal programs require many iterations for every pixel that is plotted to the screen. If you have a 486 system, these programs will really zip along, creating even the most complex fractal in 10 or 15 minutes. If you are using a more primitive machine and particularly if you don't have a math coprocessor, the programs will run a lot more slowly. You'll see the program scanning the screen, slowly plotting each pixel, in a process that may take hours. Don't get discouraged; the resulting pictures are worth waiting for. You can always stop a program before the display is completed if you want to get on with your education, but I recommend that after you get a program running, you take the time while the display is being drawn to really study the program listing until you understand the nature and purpose of each instruction.

C Programming Style

A C program contains many sections of code that are enclosed within curly brackets *{}*. I believe that the beginning curly bracket and the ending curly bracket for a particular section of code should both be indented by the same amount. Then, no matter how many subsections intervene, you can find the end of a particular section of code by simply tracing down the page from the starting curly bracket until you come to a closing curly bracket at the same level of indentation. There are many other ways of positioning the brackets, but they all tend to lead to confusion when you are trying to determine where a section of code ends in a particularly complicated program.

There are several different ways in which you can define a function. We'll get into this in more detail in Chapter 3. At this point, let's just say that the latest ANSI-approved way of defining a function, which is what will be used throughout this book, is

```
int function_name (int a, float b, char c)
```

The first word in this definition determines the type of information that is returned by the function. The second word is the name of the function. The expressions within the parentheses following the function name are pairs, each consisting of a type and a name for an argument that is passed to the function.

It is good programming practice to have a prototype for each function listed at the beginning of your program. This will be done consistently throughout this book. You can avoid this, without problems, if you make sure that every function is listed before any function that calls it. If you have a lot of intermingled functions, this can often make finding the proper order very difficult. One mistake and the compiler will bomb you out with an error message. Such mistakes cannot occur if you have the full list of prototypes near the beginning of your program.

Comments

When a compiler encounters the pair of characters /* it assumes that they and everything that follows them until and through the encountering of the character pair */ is a comment, which should be ignored by the compiler. Therefore you can fill your program with comments that improve a reader's understanding of the code. As an example, the line

```
a = 4 + b; /* b is the run number */
```

includes a comment that tells you something important about *b*, but the compiler processes only the summation and assignment, ignoring the whole comment section completely. The use of comments in a program is a good idea, since you may not be around to explain what the program is trying to do when some neophyte programmer tries to make sense it. Worse yet, years from now someone may dredge up one of your old programs and you may have totally forgotten how it works. There is a negative side to liberal commenting, however. First, if you are typing a listing into your computer, typing in the comments represents a lot of keystrokes that have to be inserted, yet these actually contribute nothing to the running of the program. Second, when you debug a program, it is often useful to comment out sections of it temporarily to simplify things while you are trying to detect errors. The /* and */ operators cannot be nested, however. Once the compiler encounters the /* operator, it takes everything after it to be a comment (including other occurrences of /*) until a */ is encountered. The very first time the */ is encountered, the compiler assumes that the comment is over. Thus, if you want to comment out a section of code successfully, you need to remove all comments that already exist within the section that you are to comment out. To make life a lot simpler for you, no comments are used with the code included in the examples in this book. It is assumed that enough descriptive material is included for each sample program to enable you to understand it without comments.

Preprocessor Directives

The term *preprocessor directives* dates back to when C compilers consisted of a preprocessor and a processor, which were run in turn to complete the compilation of a program. Most compilers are single pass now, but the terminology remains to describe particular statements which you can identify by the fact that they begin with #. Please note that preprocessor directives are not followed by a semi-colon, thus being an exception to the way that C statements in general are written. We're only going to look at two kinds of preprocessor directives here, the *#include* and *#define* directives. These are the only ones that you're likely to encounter in beginning C programs. If you get into more advanced C programming and need some other directives, refer to the instruction material that comes with your compiler.

The *#include* Directive

All of the library functions that come with your C compiler are in a group of libraries that are usually in a subdirectory called *lib*. If you are going to use any of these functions, you must have an include statement at the beginning of your program for each required library. For example

```
#include <stdio.h>
```

will make sure that the library of standard I/O functions is included in your program. You'll need to refer to the instruction material for your compiler to determine which libraries contain the functions that you are using in your program. If you neglect to include a needed library, the results can be strange and unpredictable. For example, suppose you have a statement in your program

```
a = cos(0.734);
```

If you forget to include the *math.h* library, the compiler will not give you any diagnostics to indicate this error. Instead, it will simply set *a* to 0, no matter what angle you have in the parentheses.

There are two forms of the *#include* directive. The one just shown, which includes the header file within <>, usually tells the compiler to look for the specified header file in the subdirectory *include* and for the corresponding library file in the subdirectory *lib*, both in the compiler directory. This usually will only search for the libraries that are a part of the compiler package and will not look for a directory that you have created yourself. The second form

```
#include "mylibrary.h"
```

tells the compiler to look for the header and library files in the current directory (usually the one containing the program being compiled. This is the version to use for your own libraries.

The *#define* Directive

This is the preprocessor directive that you should use to define a macro that is inserted in your code at any desired point. It is best used to create in-line substitutes for small functions, which are more convenient and faster than normal function calls. As an example, suppose you want to get the cursor position and have two functions, *getx*, which gets the *x* position, and *gety*, which gets the *y* position. You'll learn later that a function can only return one argument. Rather than getting complicated about how to return both *x* and *y* values in a single function, you can use the *#define* directive like this

```
#define getxy(x,y) {x=getx(); y=gety();}
```

Now, if you have a statement like

```
getxy(column, row);
```

it will place the *x* cursor position in *column* and the *y* cursor position in *row*.

The *#define* directive has almost unlimited power and is therefore subject to all kinds of abuse. For example, you could do this

```
#define begin {
#define end }
```

This would permit you to replace all of your curly brackets with the words *begin* or *end*, which make your programs start to look like

Pascal instead of C. With a few more *#define* statements, you can have your own version of C that is almost like Pascal and will evermore confuse anyone who tries to understand your C programs. Every year there is an obfuscated C contest in which prizes are awarded for the most confusing C programs. Unless you are entering this contest, avoid misusing the *#define* directive.

Reserved Words

Certain keywords are reserved for use by the compiler and therefore you may not use them as parameter or function names within your program. Here is a list of standard reserved words; your particular compiler may have other words reserved:

auto	enum	short
break	extern	sizeof
case	float	static
char	for	struct
continue	goto	switch
default	if	typedef
do	int	union
double	long	unsigned
else	register	while
entry	return	void

Data Types

The size of a particular data type in bits and the range of values permissible can vary from one C compiler to another, based primarily on the type of computer for which the compiler was designed. These sizes and ranges are standardized for compilers using the IBM PC and compatibles, however. Table 1-I on the next page shows the data types, sizes, and ranges for Borland C++. These are fairly representative, although a few of the more exotic types may be missing from your compiler.

Table 1-1. Data Types and Ranges

Type	Size in bits	Range
unsigned char	8	0 to 255
char	8	-128 to 127
enum	16	-32,768 to 32,767
unsigned int	16	0 to 65,535
short int	16	-32,768 to 32,767
int	16	-32,768 to 32,767
unsigned long	32	0 to 4,294,967,295
long	32	-2,147,483,648 to 2,147,483,647
float	32	3.4×10^{-38} to 3.4×10^{38} (7-digit precision)
double	64	1.7×10^{-308} to 1.7×10^{308} (15-digit precision)
long double	80	3.4×10^{-4932} to 3.4×10^{4932} (19-digit precision)

2

Writing and Compiling C Programs

In this chapter, we're going to take a look at some of the C compilers that are currently available and how they are used. Sometimes the actual mechanics of running a compiler and/or editor can get complicated and interfere with your actually getting on with C programming. This chapter will give you enough information on actual keystrokes, etc., to get you going with the program development process for whatever compiler you are using.

There are two basic ways in which C programs can be written and compiled. The first method uses what is called the *integrated development environment (IDE)*. This method starts with an editor, which you use to create a program listing that is saved in ASCII text. When you think your program is ready to run, you can tell the compiler to compile and run the code by striking certain keys. Your program is then compiled, and if the compilation is successful, the resulting compiled program is run. If there were compilation errors, the program automatically returns to the editor with the cursor positioned at the first error. You correct this and by striking certain keys advance to the next error. This process continues until you have fixed all of the errors and are ready to compile and run the program again. This is certainly the simplest way to create a C program, but it is sometimes unsatisfactory for large programs because there is not enough memory space to do everything at once.

The other method of creating a C program is through the use of the command line compiler. For this method, you use any editor that you wish to create the program listing and then save it in ASCII form to a disk file (usually ending with the extension *.c*. You then type in the name of the C compiler followed by a space and the name of your C program file on the command line (when you have a DOS prompt such as *c:>*). The simplest implementation is when your default disk and directory represent the location that contains both your compiler file and your C program file. The compiler now runs and performs the compilation. If the compilation contains errors, you will be given a listing of these errors together with the line number in the program listing where each error occurs. You are responsible for recording this list of errors and going back to your editor to find each specified line number and correcting the error on that line. You repeat this procedure until you get an errorless compilation. The compiler, at that point, creates an executable file whose name is the same as that of your program, but with the extension *.exe*. You can then run this program. We're now going to look at various available compilers in a little more detail.

Borland C++ and Turbo C++

The Borland C++, Turbo C++, and Turbo C family of compilers all use much the same integrated environment. To start the program, type *BC* if you are using Borland C++ or *TC* for one of the Turbo C family. The initial screen will come up, either empty or containing the editing window for the last file on which you were working. Type *Alt-F* to begin work on a different file. When the *File* menu window opens, use the cursor arrows to select *Open* for an existing file or *New* to create a new one. If you select *Open* a window will open that lists current files in the default directory and allows you to select one by typing in a name or using the cursor arrows. If you select *New* a blank editing window will appear headed by the file name *NONAMEnn.C*, where *nn* is a two digit number beginning with 00. (You can rename this file to whatever you want when you save it.) After you are through typing in or editing your file, type *Alt-R* and select *Run*. Your program will then be compiled. A window will keep track of how many lines are being compiled and how many warnings and errors there are. If there are any errors, a blinking legend will say *Press any key*. When you press a key, the editing window will return, with the line containing the first error or warning highlighted. Hitting the *Ent* key will allow you to edit the program to correct this error. Then typing *Alt-F8* (the F8 special key, not the separate keys for F and 8) will move you down to the next error or warning. When you have finished your editing process, you can hit *Alt-R* to recompile. When the program no longer contains errors, it will automatically be run after compilation is finished.

Borland C++ and Turbo C++ are available from

Borland International
P. O. Box 660001
Scots Valley, CA 95067-0001
Telephone: (408)438-8400

If you're only interested in learning C right now, don't despair that these compilers are both for C++. Either compiler will revert to a pure C compiler if you use the extension *.c* on your source code program file name. Later, if you decide to get into C++, you can automatically invoke the C++ compiler by using the extension *.cpp* on your file name. Borland C++ includes a number of tools for generating programs that run from Windows; Turbo C++ does not. If you are going to be doing a lot of heavy-duty C language development, or if you are going to be programming for use with Windows, you'll need Borland C++. Otherwise, you can probably get along with Turbo C++, which has most of the features and is much cheaper. One valuable feature of Borland C++ is the ability to use extended memory in its compiling process, making it possible to handle much larger programs than can be handled with Turbo C++.

All of the programs in this book were developed and tested using Borland C++, version 3.0.

Zortech C++

Like Borland C++ and Turbo C++, Zortech C++ includes an excellent compiler for C, which you can use to compile all the programs in this book. Zortech's version of the IDE is called the Zortech Workbench. You start this program by typing *ZWB*. The program will begin by indicating *Read file:*, at which point you should enter the name of the file for the program that you are writing. The listing of your program should now appear on the edit screen. You can edit or add to it as necessary. When you are ready to compile and run the program, type *Alt-C*. The *Compile* menu will now come up on your screen, giving you several options that can be selected with the cursor arrows. Zortech doesn't have the *Compile* and *Run* functions tied together, so don't select *Run Program* yet, since you don't have a compiled version of the program. (If you attempt to run a nonexistent program, the workbench will go off to never-never land and you will have to reboot.) For simple programs like those in this book, you should select the *Compile* option. The program will chug away for awhile and then return you to the edit screen, either with some errors to be fixed or with a notation *Compiled successfully.* Now you can type *Alt-C* again and this time select the *Run Program* option. The program will now ask for *Program command line: (include program name).* Type in the name that you assigned to your program, without any extension. The program should now begin to run. When the program is complete, you will get a blank screen with the legend *Hit any key to proceed.* When you do this, you will be returned to the edit screen. Although my test program ran just fine, when I got back to the edit screen, there was the message *Process terminated with an error.*

Zortech C++ is available from

Zortech, Incorporated
1165 Massachusetts Ave.
Arlington, MA 02174
Telephone: (617)646-6703

The version I used in my testing was version 2.12.

Microsoft C++

Like Borland C++ and Turbo C++, Microsoft C++ includes an excellent compiler for C, which you can use to compile all the programs in this book. MicroSoft's version of the IDE is called the Programmer's Workbench. You start this program by typing *PWB*. The program will begin by coming up with the last program listing that you were working on, or an empty screen if there was no previous program. After you are through editing or adding to your program, type *Alt-R*. The *Run* menu will now come up on your screen, giving you several options that can be selected with the cursor arrows. Choose the *Execute:* option. Another menu will appear that asks *Do you want to Build/Rebuild Current Target?* Select option *B* to compile the program. Microsoft C++ will run for awhile to compile and link your program. At the end of that process another window appears on the screen which tells you that the operation is complete and indicates how many errors and warnings occurred and gives you several choices for the next action. If the compilation was successful, you may select *R* to run it; if there were errors, you can return to the editor and correct them. Once the program has run, you are returned to the edit screen.

Microsoft C++ is available from

Microsoft Corporation
One Microsoft Way
Redmond, WA 98052-6399
Telephone: (206)454-2030

Microsoft C++, version 7.0, is the first version of Microsoft C to include the additional C++ features. The version I used in my testing was version 6.0, which is a C only compiler.

Power C

If you can get along without the IDE, Power C from Mix Software is about the best buy around. Power C costs $19.95 and comes with a 664-page instruction manual that is one of the better tutorials on C. To compile and run a program from the command line, you type *PC* followed by the file name of your program. I did this with one of the programs that had been tested with the other compilers just described and it compiled and ran with absolutely no difficulty. In fact, the Power C compiler appears to produce more efficient code than some of the other compilers, because the resulting program appeared to run faster. Power C is available from

Mix Software, Inc.
1132 Commerce Drive
Richardson, TX 75081
Telephone: (214)783-6001

3

Main Program and Functions

You are now ready to learn the basic elements of programming in C. In this chapter you'll write simple programs that aren't very exciting, but they do demonstrate the principles of using the C language. Once you have mastered these principles, you'll be ready to expand to more complex and exciting programs.

Before you begin programming in C, you need to have a fundamental understanding of the way in which C programs are structured. A C program is constructed from building blocks called *functions*. A function performs some particular set of programming actions. It can be called as often as needed. Every C program consists of one or more *functions*. A *function* consists of a set of curly brackets which contain a series of statements, each ending in a semicolon. The curly brackets tell the compiler to handle whatever is within them as a single unit, but note that the closing curly bracket is not followed by a semicolon as would be the case if it was a single statement.

The *main* Program

Each C program must have one and only one *function* called *main*. Execution of the program begins with the start of the *main* function and ends when it ends. The simplest possible C program looks like this

```
void main(void)
{
}
```

Not only is it simple, it doesn't do anything except tell the computer to start and then stop running the program. Now we'll try a program that is a little more complicated.

```
void main(void)
{
	printf("\nThis is my first C program...");
}
```

This program will display the sentence *This is my first C program...* on your display screen. (We're going to describe the *printf* statement in Chapter 7, so don't worry about it right now.) The first important thing to look at is the makeup of the first line, which defines the *main* function. A function begins with a header. The first word of the header describes the type of data that is returned by the function. This may be any type of data defined by the compiler or by you in a *typedef* statement. Since this program ends when *main* ends, nothing is ever returned, so *main* should always be of type *void* which means that nothing is returned. (You can often get away with leaving this word out for the main program, but some compilers will then issue a warning saying that you should return something, so it's best to include the *void*.) The next word in the function description is the name of the function, which in this case is *main*.

Most functions can have as many arguments as you wish enclosed within a set of parentheses following the program name. The program *main* is different, however. By ANSI definition, it may have either 0 or 2 arguments. The 0 argument case is shown above, where the type *void* is enclosed within the parentheses. The two-argument case looks like this:

```
void main (int argc, char *argv)
```

The *argc* and *argv* arguments allow you to transfer information from the command line that was used to execute the program to the program itself. When you execute the program you enter the name of the program followed by any necessary arguments, separated from the program name and from each other by one or more spaces. This data is stored in the command line buffer. The first (*argc*) argument is an integer which is the number of arguments that you entered on the command line, including the program name. The second (**argv*) argument is the address of the start of an array of null terminated strings, each containing one of the arguments from the command line.

(The *argv[0]* string is the name of the program being executed.) You may use *argc* to determine how many arguments you are going to read from the array and how to use them and may also use it to produce an error message and terminate the program if the proper number of arguments was not entered on the command line. You can read each argument in the array, convert it to a number, or whatever you want to do with it. We'll get into how to do this in Chapter 14. Most of the programs you'll be using in this book, however, won't include any command line arguments.

Functions within a Program

You may be wondering why a C program doesn't always consist of just the function *main*, made up of a long list of statements. There are only two reasons for having additional functions within a program. The first is that you have a section of code that needs to repeated at several different points within your program. It's much simpler and more compact to have this section of code defined as a separate function and then call it each time it's needed, rather than to repeat the code at each point where it is used. The second reason for having other functions is to separate a very long and complicated program into sections that are more easily understood. (Also, it is possible to have a program so long that the compiler can't handle it all in one chunk so that you have to break it up into separate functions.) It is very bad programming practice to have a lot of short functions when they are not really needed. This causes anyone who is trying to read and understand the program listing to have to jump back and forth continually and makes understanding much more difficult. So, when you are writing a C program, look with suspicion upon any function that is not used more than once in the program. If there isn't a very good reason for having it be a separate function, incorporate the code into the main function.

With these introductory comments, let's look at exactly what a function can and cannot do. Here is a very simple function to illustrate.

```
int a, b, c;

int add (int x, int y);

void main (void)
```

```
{
        a = 3;
        b = 4;
        c = add(a,b);
        printf("\nc = %d",c);
}

int add(int x, int y)
{
        int z;

        z = x + y;
        return (z);
}
```

First, look at the header of the function *add*. The first word, *int*, shows that the information returned by this function is of type *int* (an integer). Next comes the function name, *add*. Next, the information in the parentheses tells us that two arguments, x and y, are passed to the function, each of which is an integer. Note that the argument names specified in the header are the names that these arguments take within the function itself. The arguments that you use when you call the function will determine what is passed to it from the calling function. Now look at the body of the *add* function. First it defines a new integer variable, z. It then adds x and y to obtain z. Then it returns the value of z. Now let's go back to the *main* program. At the very beginning we have a prototype for the function *add*. This is the same as the header for the function except that it is terminated by a semicolon. Modern practice requires that at the beginning of a program you have a prototype for every function used in the program. This is needed so the compiler will know the characteristics of the function when it first encounters a call to it. If you don't do this, the compiler will make certain assumptions about the function when the first call is received, which may not be what you want it to assume. For example, if the compiler encounters a call to a function when it has not yet seen a prototype or the function definition, it will assume that the type is *int*. If you then define the function to be of a different type, you'll get an error message and the compilation will bomb out.

Next, the *main* program defines two integers, a and b. Now, c is set equal to the *add* function and the arguments a and b are passed to it. What occurs is that copies of the values a and b are passed to *add*. The function then does its thing and adds together the two arguments ($3 + 4 = 7$) and returns the 7, which is then stored in c. The program then displays this value on the screen.

Function Limitations

The principal limitation in using a function is that the function can only return one argument. There are many times when we would like the function to perform operations that change several variables. There are a couple of ways that we can work around the basic limitation of the function. The first method involves pointers. It will be covered in complete detail in Chapter 10. Here, we'll just give a cursory summary. There are two ways of describing an argument in C. If we have an argument that is defined as

```
int a;
```

then when C performs operations upon *a*, it is manipulating the value of this argument. For such an argument, *&a* is the address where the argument *a* is located. Another possibility is to have the definition

```
int *b;
```

In this case, the asterisk indicates that *b* is the address where an argument is located. Now, if the program contains an expression like

```
b = c;
```

it means that we are changing the address currently in *b* to whatever value is in *c*. If we want to change the value of the argument pointed to by *b*, we must have an expression like

```
*b = c;
```

which says that we replace the contents of the memory address pointed to by *b* with the value of *c*. Now let's see how this applies to a function call. Suppose we have these definitions

```
int a;
void function_name(int *b);
```

This says that we are passing an address to the function. (Remember that only a copy is actually passed.) Now, somewhere within our program we have a function call

```
function_name(&a);
```

This works just fine. We have *a* defined as an argument and we pass the address of *a* (as given by *&a*) to the function where the address becomes known as b. Now, within the function, we have a statement like

```
*b = 6;
```

This says that we replace the contents of the address b with 6. But you remember that in the outside world, when we leave this function, that address is known to contain the value of a, so that when we return from the function and look at a, we find that it contains the new value 6. In this way, we can pass several address pointers to a function, change the values contained at these addresses within the function, and have the new values at the appropriate addresses when we return. What makes the situation confusing is that when we define a name, having it preceded by an asterisk means that it is a pointer to an address, whereas an asterisk preceding a name elsewhere in a program means that we are going to the address indicated by the name and perform the specified operations on its contents.

The other way of getting around the limitation of a function only returning a single variable is through the use of the *typedef* and *struct* statements. We already have such data types as *char, int*, and *float* defined as part of the C language. The *typedef* statement allows us to define a new data type. If we define it as a *struct*, it may contain several pieces of data, which are considered as a single entity. This is described in detail in Chapter 11. As an example, suppose we have

```
typedef struct
{
      int x;
      int y;
} POSITION;

POSITION curs;
```

Now if we have a function defined as

```
POSITION getxy(void)
{
      POSITION xy;
      xy.x = 6;
      xy.y = 4;
      return xy;
}
```

we can have a function call in our program like

```
curs = getxy();
```

and after execution, we will find that in our main program *curs.x* is 6 and *curs.y* is 4.

Summary

You should now have a feeling as to how to write a main program and how to include functions, including those which need to return more than one argument. Next we take an excursion to see how the C language can interact with DOS and ROM BIOS functions in a PC. Once we have covered that topic, we'll be ready to write some real programs that illustrate new capabilities of the C language and produce some interesting fractal displays.

4

Interaction of C with DOS

DOS Fundamentals

The basic operating system of your PC computer is some version of the Disk Operating System (DOS). DOS consists of two parts. The Basic Input/Output System (BIOS) is permanently burned into chips on your motherboard and peripheral boards. (DOS is written so that upon booting up, a number of dedicated memory addresses are checked, and if they contain the proper header, the following content is added as part of the BIOS. This permits special additions to BIOS to exist for graphics cards, serial/parallel I/O cards, disk controller cards, etc. Thus the BIOS is really customized for the particular cards that you have in your system.) All of the various operations of DOS that are accessible from the outside world are performed through setting registers with particular values and then calling an interrupt. We do not go into detail on what DOS services are available; if you're going to get into the subject in depth you need to get one of the books that specializes in this field. The purpose of this chapter is to acquaint you with the fact that you may access any of the DOS services from your C program and to show some simple examples that will be used repeatedly through the rest of the book.

Using DOS from C: the *int86* Function

The most commonly used way of communicating with a DOS interrupt from C is through the use of the *int86* function. This function takes the form

```
int int86(int intno, union REGS *inregs, union REGS
      *outregs);
```

Before using this function, you have to define one or two variables of the type *union REGS*. Chapter 11 will go into detail on how structure and unions work. Here, we'll simply say that this union contains data for the *a, b, c,* and *d* registers of the microprocessor, and permits you to specify this data either on a byte or word basis. If you just define one of these variables, you will have to use it both for the input and output register data, so that after you call the *int86* function, your variable will contain the new register contents and any old contents will be lost. Alternately, you can define separate variables for input and output and thereby preserve both sets of values. In using the *int86* function, you first set up the register contents as needed and then call the function. In the sections that follow, we're going to provide some simple examples. Bear in mind that you can extend this technique to call any DOS function that you desire.

The *setmode* Function

The *setmode* function uses the ROM BIOS video service of DOS to set the video mode of the display system. The function is called from C like this

```
/*
          setmode() = Sets the video mode
*/

void setmode(int mode)
{
      union REGS reg;

      reg.x.ax = mode;
      int86 (0x10,&reg,&reg);
}
```

The ROM BIOS video service is called through interrupt number 10H, so that is the number that goes into the first parameter of the *int86* function call. We define the union REGS parameter structure *reg* to hold the register contents and use it for both the input and output registers. (In this case the function returns nothing so we aren't really interested in the output.) To make the display mode change, we need to put the number 00 into the upper byte of the *a* register and the desired mode into the lower byte of the *a* register. This is accomplished by the line

```
reg.x.ax = mode;
```

The mode that we will use throughout the book is mode 18, which sets up the display screen for the graphics mode having 640 × 480 pixels × 16 colors. Most compilers will get you back to the text mode after you are through running one of the fractal displays that will be given later. If this turns out not to be the case, you need to close each program with the statement *setmode(3);* to return to the normal text mode. The fractal example programs found throughout the book assume that you have a VGA display adapter and monitor. If your display system is an EGA, you'll need to use graphics mode 16 and change various references to 480 throughout each program to 350.

The *cls* Function

The next function that we're going to use a lot is one to clear the graphics screen to a selected color. In most of the sample programs, you won't really need to clear the screen, since the *setmode* function does the initial clearing. However, *setmode* leaves the screen black. A number of the sample programs take the longest to fill in black areas of the picture; if the screen is black to start with, you won't be able to see what progress the program is making in painting in black areas, which can be rather frustrating. Therefore we usually clear the screen to a light gray, which makes progress in painting black areas obvious. The function to do this is

```
/*

        cls() = Clears the screen to a selected color

*/

void cls(int color)
{
        union REGS reg;

        reg.x.ax = 0x0600;
        reg.x.cx = 0;
        reg.x.dx = 0x1E4F;
        reg.h.bh = color;
        int86(0x10,&reg,&reg);
}
```

Again the interrupt 10H is used. To do the screen clearing job, we use the ROM BIOS video service that scrolls the screen up. This is accessed by having the high-order byte of register *a* set to 6 and the low-order byte to 0. The high-order byte of register *c* contains the row number for the upper left corner of the window being scrolled and the low-order byte of this register contains the column number for the upper left corner. In this case, we're working with the entire screen, so both of these are 0. In a similar manner, register *d* contains the values for the lower right corner of the window. We're talking in terms of characters rather than pixels; for the graphics mode that we're in, the row number of the lower right corner is set to 1EH (30 decimal) and the column number is 4FH (79 decimal). The row number for the lower right corner doesn't have to be exact. However, if it's too small, you won't color the entire screen and if it's much too large, you'll slow up the screen clearing process and may get into some memory area that messes up your display or program. Finally, you have to put the desired color (0 to 15) in the high-order byte of register *b*. You're then ready to make the call to *int86* to clear the screen.

The *plot* Function

The final function that we're going to use regularly is *plot*, which plots a single point to the screen at a designated location (in row and column pixels) and in a selected color. The function to do this is

```
/*

        plot() = Plots a point on the screen at a
             designated position using a selected
                  color for 16 color modes.

*/

void plot(int x, int y, int color)
{
      #define graph_out(index,val)  {outp(0x3CE,index);\
                                          outp(0x3CF,val);}

      int dummy,mask;
      char far * address;

      address = (char far *) 0xA0000000L + (long)y * xres/8L
            + ((long)x / 8L);
```

```
        mask = 0x8000 >> (x % 8);
        outport(0x3CE,mask | 8);
        outport(0x3CE,0x0205);
        dummy = *address;
        *address = color;
        outport(0x3CE,0x0005);
        outport(0x3CE,0xFF08);
}
```

This function is a little different from the others in that it uses calls to output ports and direct memory addressing to send the pixel to the screen. If you're not interested in the nitty-gritty details, just skip the rest of this section. In the 16 color modes, the VGA stores the color information for a pixel in four different memory planes. These planes are all accessed from the same computer memory address and internal circuitry on the VGA card sorts out which color information goes to each plane. To start the process, we determine the memory address of the byte where this particular pixel is stored. (Since 8 pixels are stored in each byte, we need to divide the column number by 8 and then add 80 bytes per row to it to find the offset from the base address.) We then set up *mask*, which allows only the selected pixel in the byte to be changed. We then output the proper setting to the VGA ports to allow the selected pixel in each memory plane to be changed. Next, we do a dummy read of the selected memory address. This causes the data in each of the four memory planes to be latched into internal VGA registers. Then the color information is sent to the selected memory address. The VGA modifies the selected pixel in each memory plane as necessary so that the new color is displayed at the desired location. Then all we have to do is restore the VGA registers to their default settings. This is just a cursory description of the operation. If you want to get into the subject in detail, see *The C Graphics Handbook*.

Summary

This has been just a quick introduction to how you can make your C programs interact with available DOS services. You can get a book on DOS and find a lot of other things that you can do if you want to get into this farther, but for right now, you'd be better off to concentrate on learning the C language as described in the chapters that follow. You may wonder why I've introduced the subject at all. First, this is a very important area of programming that is often totally ignored in beginning C books and I think you need to at least know that it exists. Second, the simple functions given in this chapter enable you to create fractal graphics displays that will work with any C compiler that uses a PC, rather than having to try to modify each program to use the graphics capabilities of your particular compiler. Finally, these functions serve to dispel some of the mystery that surrounds the use of graphics functions. You can see that a lot of highly unusual graphics output can be produced with a few relatively simple functions.

5

Operators and Expressions

One more thing that we have to do before actually getting into C programming is look at the operators used in C. These operators look a lot like those used in ordinary mathematical expressions, so you shouldn't have too much trouble writing equations using variables that you have defined together with the C operators. For example,

```
int a, b, c, d, e, f;

c = a + b;
d = b - a;
e = b / a;
f = b * a;
```

You start by defining the integers *a, b, c, d, e* and *f*. Following this, the first three expressions look just like ordinary mathematical expressions and specify an addition, a subtraction, and a division, respectively. In interpreting the last expression, you need to know that the operator * stands for multiplication in a C mathematical expression. Then the meaning of this expression becomes obvious. The only thing you need to remember is that you have defined all the arguments as integers. Therefore, when you perform a division, you end up with an integer; if there is a fractional part of the division result, it is truncated. [If you need the fractional part, you should have defined the arguments as floating point numbers (type *float*)]. Then the same operators would perform floating point operations instead of integer ones. C always keeps track of the type of data that you are using and performs the correct version for the data type of each operation.

There are a few operators that are less obvious in their usage. We list them all in Table 5-I and provide some additional description so that you'll know how they work.

Operator Definitions

Table 5-I lists the C operators and describes their usage. You may need to refer to this table when you are trying to figure out from a program listing what the program is doing, or when you are writing a program and want to perform some arcane action.

Table 5-I. C Operators and Their Definitions

Operator	Operation
[ ]	Brackets enclose array subscripts. Example: `int a[40];` sets up an array of 40 integers. `b = a[22];` assigns the value of `a[22]` to `b`. (Note that since an array begins with the 0th member, `a[22]` is the 23rd value in the `a` array.)
( )	Parentheses group expressions, isolate conditional expressions, indicate function calls and parameters.

{ }	Braces enclose a compound statement. Example:
	`if (r == 3)`
	`{`
	`    a = 3;`
	`    b = 6;`
	`}`
	When the condition in the *if* statement is true, both statements within the braces are performed.

.	Direct component selector for a member of a structure or union. Example: `complex.r` where `complex` is a structure or union.
->	Indirect component selector for a member of a structure or union. Example: `complex->r` where `complex` is a pointer to a structure or union.

(continued)

Table 5-I. C Operators and Their Definitions (cont.)

Operator	Operation
++	Increment. `b = ++a;` means that *a* is incremented before the other operations take place. `b = a++;` means that *a* is incremented after the other operations take place. Thus if *a* is 3 then for the first type operation *b* is 4 and for the second type *b* is 3. In either case *a* is 4 after the statement has been executed.
--	Decrement. `b = --a;` means that *a* is decremented before the other operations take place. `b = a--;` means that *a* is decremented after the other operations take place. Thus if *a* is 4 then for the first type operation *b* is 3 and for the second type *b* is 4. In either case *a* is 3 after the statement has been executed.
&	`a & b` means that *a* and *b* are bitwise ANDed together. `&name` means that the address of *name* is taken.
*	`a * b` means that *a* is multiplied by *b*. `*ptr` means that the contents of the address given by *ptr* is taken.
+	`+ a` means that *a* is a positive number. `a + b` means that *a* and *b* are added together.
-	`- a` means that *a* is a negative number. `a - b` means that *b* is subtracted from a.
~	Bitwise 1's complement.
!	Logical negation.
sizeof	`sizeof(b)` takes the size of *b* in bytes.
/	Division.

(continued)

Table 5-I. C Operators and Their Definitions (cont.)

Operator	Operation
%	Remainder. This is the result of modulus arithmetic. Example: `c = a % b;` means that *a* is divided by *b* and the remainder is assigned to *c*. If *a* is 17 and *b* is 3, *c* would be equal to 2.

<<	Shift left. The binary representation of a character or integer is shifted to the left the number of places indicated. Example:
	`int a = 35;`
	`int c;`
	`c = a << 3;`
	The binary equivalent of 35 is 100011. Shifted 3 places to the left this gives 100011000 which is equivalent to decimal 280 so 280 would be assigned to *c*.

>>	Shift right. The binary representation of a character or integer is shifted to the left the number of places indicated. Example:
	`int a = 35;`
	`int c;`
	`c = a >> 3;`
	The binary equivalent of 35 is 100011. Shifted 3 places to the right this gives 100 which is equivalent to decimal 4 so 4 would be assigned to *c*.

<	Relational less than. Example: `if (a < b)` is true and the statements below the *if* are executed if *a* is less than *b*.
>	Relational greater than. Example: `if (a > b)` is true and the statements below the *if* are executed if *a* is greater than *b*.
<=	Relational less than or equal to. Example: `if (a <= b)` is true and the statements below the *if* are executed if *a* is less than or equal to *b*.

(continued)

Table 5-I. C Operators and Their Definitions (cont.)

Operator	Operation
>=	Relational greater than or equal to. Example: `if (a >= b)` is true and the statements below the *if* are executed if *a* is greater than or equal to *b*.
==	Relational equality. Example: `if (a == b)` is true and the statements below the *if* are executed if *a* is or equal to *b*.
!=	Relational inequality. Example: `if (a != b)` is true and the statements below the *if* are executed if *a* is not equal to *b*.
^	Bitwise XOR (exclusive OR). `a & b` means that *a* and *b* are bitwise exclusive ORed together.
|	Bitwise OR. `a & b` means that *a* and *b* are bitwise ORed together.
&&	Logical AND. Example `(if (a == 3) && (b == 4))` means that the condition is true if both *a* is equal to 3 and *b* is equal to 4.
||	Logical OR. Example: `(if (a == 3) || (b == 4))` means that the condition is true if either *a* is equal to 3 or *b* is equal to 4.
?:	Question mark and colon (separates parts of trinary conditional statement). Example: `c = a < b ? 3 : 4;` means that if *a* is less than *b* then *c* is set equal to 3 and otherwise it is set equal to 4.
=	Equals sign. `a = b;` assigns the value of *a* to *b*. As an initiator `int a[3] = {3,4,5};` means that array members *a[0]*, *a[1]* and *a[2]* are initiated to values of 3, 4 and 5, respectively.

(continued)

Table 5-I. C Operators and Their Definitions (cont.)

Operator	Operation
*=	The value on the left side is assigned the value of the left side multiplied by the value of the right side. Example: `c *= a;` means that c is assigned a new value that is equal to the old value of c multiplied by a.
/=	The value on the left side is assigned the value of the left side divided by the value of the right side. Example: `c /= a;` means that c is assigned a new value that is equal to the old value of c divided by a.
%=	The value on the left side is assigned the value of the remainder when the value of the left side is divided by the value of the right side. Example: `c %= a;` means that c is assigned a new value that is equal to the remainder when the old value of c is divided by a.
+=	The value on the left side is assigned the value of the left side added to by the value of the right side. Example: `c += a;` means that c is assigned a new value that is equal to the old value of c added to a.
-=	The value on the left side is assigned the value of the right side subtracted from the value of the right side. Example: `c -= a;` means that c is assigned a new value that is equal to the value of a subtracted from the old value of c.
<<=	The value on the left side is assigned the value of the left side shifted left by the number of bits specified by the right side. Example: `c <<= 3;` means that c is assigned a new value that is equal to the old value of c shifted left 3 bits.

(continued)

Table 5-I. C Operators and Their Definitions (cont.)

Operator	Operation
>>=	The value on the left side is assigned the value of the left side shifted right by the number of bits specified by the right side. Example: `c >>= 3;` means that *c* is assigned a new value that is equal to the old value of *c* shifted right 3 bits.
&=	The value on the left side is assigned the value of the left side bitwise ANDed with the value of the right side. Example: `c &= a;` means that *c* is assigned a new value that is equal to the old value of *c* ANDed with *a*.
^=	The value on the left side is assigned the value of the left side bitwise XORed (exclusive ORed) with the value of the right side. Example: `c ^= a;` means that *c* is assigned a new value that is equal to the old value of *c* XORed with *a*.
|=	The value on the left side is assigned the value of the left side bitwise ORed with the value of the right side. Example: `c |= a;` means that *c* is assigned a new value that is equal to the old value of *c* ORed with *a*.
,	Comma (separates elements of function argument list)
;	Semicolon (statement terminator)
:	Colon (indicates labeled statement)
...	Ellipsis (indicates variable number of arguments)
#	Identifies a preprocessor instruction.

Operator Precedence

If you begin to create complex expressions, you're going to be concerned with the precedence of operators. Take a look at the following program to see if you can determine the problem:

```
#include <stdio.h>

void main(void)
{
        int a, b, c, d;

        a = 3;
        b = 8;
        c = 2;
        d = a + b / c;
        printf("\na: %d   b: %d  c: %d   a+b/c: %d",
              a,b,c,d);
        d = (a + b) / c;
        printf("\na: %d   b: %d  c: %d(a+b)/c: %d",
              a,b,c,d);
        getch();
}
```

There is no trouble understanding how the second evaluation of *d* is obtained. First *a* and *b* are added together to obtain 11 and then this is divided by *c* to obtain 5. The first evaluation of *d* is more obscure. It could work just the same as the second evaluation. Alternately, we could first divide *b* by *c* to obtain 4 and then add this to *a* to obtain 7. It all depends on what precedence C assigns to the operators. If it does additions first and then divisions, we'll obtain a result of 5; if it does divisions first and then additions, we'll obtain a result of 7. Whether C starts at the left and works toward the right or starts at the right and works toward the left may also be important. You can always use enough sets of parentheses to make your equation totally unambiguous, but you'll probably want to eliminate any parentheses that aren't absolutely necessary so as to keep your code as simple as possible. Table 5-II shows the precedence for the various operators and the direction used in their evaluation. The highest precedence is 1; the lowest is 15. This isn't the exact order or direction that C uses in its actual compiled code; the compiler does a lot of rearranging to produce the most efficient code. It does follow the rules, however, so that if you use the table to determine what's going to happen with an apparently ambiguous section of code, you'll be all right. If you have any question about whether you are doing things right, run a small test program to make sure that the result is what you expect.

Table 5-II. Precedence of C Operators

<table>
<tr><th>Operators</th><th>Precedence</th><th>Direction</th></tr>
<tr><td>( ) [ ] -> .</td><td>1</td><td>Left to right</td></tr>
<tr><td>! - + - ++ -- b & (unary operators) * (indirection operator) (typecast) sizeof</td><td>2</td><td>Right to left</td></tr>
<tr><td>* / %</td><td>3</td><td>Left to right</td></tr>
<tr><td>+ -</td><td>4</td><td>Left to right</td></tr>
<tr><td><< >></td><td>5</td><td>Left to right</td></tr>
<tr><td>< <= > >=</td><td>6</td><td>Left to right</td></tr>
<tr><td>== !=</td><td>7</td><td>Left to right</td></tr>
<tr><td>&</td><td>8</td><td>Left to right</td></tr>
<tr><td>^</td><td>9</td><td>Left to right</td></tr>
<tr><td>|</td><td>10</td><td>Left to right</td></tr>
<tr><td>&&</td><td>11</td><td>Left to right</td></tr>
<tr><td>| |</td><td>12</td><td>Left to right</td></tr>
<tr><td>? : (trinary conditional operators)</td><td>13</td><td>Right to left</td></tr>
<tr><td>= *= /= %= -= &= ^= |= <<= >>=</td><td>14</td><td>Right to left</td></tr>
<tr><td>,</td><td>15</td><td>Left to right</td></tr>
</table>

Mixed Types and Typecasting

What happens if you write a C expression in which different types of data are involved? C will automatically change data types, but only when absolutely necessary. This can sometimes give unexpected results. This program will demonstrate all that you ever need to know about using mixed data types and typecasting.

```
#include <stdio.h>

void main(void)
{
        int a, b, c;
        float d, e;

        a = 3;
        b = 8;
        d = 8.0;
        e = b / a;
        printf("\ne: %f", e);
        e = (float)b / a;
        printf("\ne: %f", e);
        e = d / a;
        printf("\ne: %f", e);
        getch();
}
```

The result of running this program is:

```
e: 2.000000
e: 2.666667
e: 2.666667
```

Let's see what this means. The expression *e = b / a;* says that the integer *b* is divided by the integer *a* and the result placed in the floating point number *e*. C doesn't do any type conversion to divide one integer by another so 8 is divided by 3, giving the integer value 2. This is then converted to a floating point number (2.000000) and stored as *e*. (If you wanted to get 2.666667, you're unpleasantly surprised.) Next, look at the expression *e = (float)b / a;*. A C data item can be typecast by putting the type name in parentheses, *(float)* in this example, before the data item (*b*). In evaluating the expression *b* is changed to a floating point number (8.000000). Now C must perform the division of a floating point number by an integer, so it automatically converts the integer to floating point and does the division, getting 2.666667, which is stored as *e*. To get the proper result, you don't have to typecast both *a* and *b*. Typecasting either one will cause C to convert the other automatically. Finally we turn to the expression *e = d/a;*. Since *d* is a floating point number, C will automatically convert *a* to a floating point number and the result in *e* will again be 2.666667. You can learn from this example that you should typecast as many data items as you think necessary when you are in doubt about how C is going to process an expression containing mixed data types. It doesn't hurt to typecast items unnecessarily. Better to do excessive typecasting than get a result that you don't want.

6

Initializing Variables

Global and Local Variables

Every variable used in C has to be defined as to its type, either at the beginning of the program for global variables or at the beginning of a function for local variables. A global variable, which is defined before the main or any other function, may be accessed by any function in your C program. A local variable may only be accessed by the function in which it is defined. Now, suppose you have a global variable *x* and within one of your functions you also define a local variable having the same name, *x*. The local variable will be set up within the function and any references to *x* that you make within the function will be to the local version. The global variable *x* will have the same value when you leave the function that it had when you entered it; no changes to the local *x* will have any effect on the global variable of the same name. Finally, no matter what value the global variable *x* had when you entered the function, the initial value of the local *x* will be whatever you initialize it to, or if you don't initialize it, it will be initialized to 0 automatically.

Initializing Data Items

Variables that you plan to use later in a program or function are defined by such statements as

```
int a, b, c, d, e;
char ch, ch1, ch2, ch3;
float x, y, z;
```

When these statements are encountered, the names of the variables will be assigned by C, space in memory will be allocated sufficient to hold the value of each variable, and each variable will be initialized to 0. Suppose 0 isn't the value that you want to begin with? Then you can initialize variables in the definition statement like this

```
int a=3, b, c=4, d=0x4345, e='gh';
char ch=0x43, ch1='\43', ch2='c', ch3=22;
float x=3.14159, y=2.718, z;
```

You can initialize an integer (which in most PC C compilers is 2 bytes) with a number, or with a hexadecimal number preceded by *0x*, which is 2 bytes long, or by pair of ASCII characters within single quotes. Within the list of integers there may be one or more that aren't initialized at all and therefore will take the default value of 0. You can initialize a character (which is one byte long for the PC) with a hexadecimal number preceded by *0x* or by a hexadecimal number preceded by a \ within single quotes or by an ASCII character within single quotes or by a number, or not at all. You can initialize a floating point number with a decimal number. (You can probably initialize a floating point number with a 4-byte hexadecimal, but it's hard to say just what the program would make of this and there is really no reason why you would want to.)

Initializing Arrays

Things get a little more complicated when you're initializing arrays, particularly when the arrays are multidimensional. This program demonstrates some of the complications

```
#include <stdio.h>

void main(void)
{
      int i,j,b[40] = {1, 2, 3, 4, 5} ,
            a[8][6] = {{1,2,3,4},{0,5,6,7,8,9},{0},
            {12,13,14,15,16,17}}, c;
      char ch[40] = {"Now is the time..."},
      ch1[40] = {'T','h','i', 's',
            ' ', 'i', 's', ' ', 'i', 't'},
      ch2[3][40] = {{"This is line 1..."},
            {"This is line 2..."},
            {"This is line 3..."}};
      printf("\n%d  %d  %d  %d  %d  %d", b[0],
            b[1], b[2], b[3], b[4], b[5]);
      printf("\n%s",ch);
      printf("\n%s",ch1);
      for (i=0; i<8; i++)
```

```
        {
            printf("\n");
            for (j=0; j<6; j++)
                printf("%2d   ",a[i][j]);
        }
        printf("\n%s\n%s\n%s",ch2[0], ch2[1],ch2[2]);
        getch();
}
```

First, look at array *b*. It contains 40 members, but we have only initialized the first five. (The remaining ones are filled with 0s by default.) All of the initializing values are enclosed within curly brackets and are separated by commas. These go into the array in order starting with the 0th member. If you wish to skip some members you must identify them with 0's in the list. The *printf* statement displays the first six members of the array, which includes the first five initialized values and a sixth (0) value.

That was fairly easy. Now let's look at an array of characters, *ch*. We can initialize it by a text string enclosed within double quotation marks which are in turn enclosed in curly brackets. If you want to treat this array as a string, you have to make sure that there is at least one more member in the array than the number of characters that you send to it. This is because the last member of a string must be a NULL (0) character. Initializing the way we just did will automatically put the NULL at the end if there is enough space. Another way of initializing a character array is shown for the array *ch1*. In this, the curly brackets enclose a number of individual characters, each enclosed in single quotation marks and separated by commas. In this case the NULL won't be automatically inserted at the end, but if you have an extra member there, it will be set to 0 (NULL) by default. This works just as well as the previous method, but is a little lengthier. You might want to use it if you are displaying a number of special-purpose characters. Two *printf* statements in the program display the two strings.

Next we are going to initialize a two-dimensional array, *a*. Note that the order of initialization is that we begin with the first array dimension set to 0 and initialize for all members of the second dimension. Then, the first dimension is set to 1 and all members of the second dimension are initialized, and so forth. This is done by having one set of curly brackets that encloses all of the initialization values, and within these brackets another set of curly brackets for each value of the first dimension that is to be initialized, separated by commas. You have to be careful with a few things here. For the first dimension set to 0, we initialize the first four members of the second dimension. The remaining two are set to 0 by default. For the first dimension set to 1, we want to initialize all members of the second dimension except the 0th. However, we must set this to 0 to get the values into the proper positions.When the first dimension is 2, we don't want to initialize any of the members. However, we need to have a set of curly brackets for this dimension, and it must include at least one 0. For the first dimension set to 3, we initialize all six members of the second dimension. This is the end of the initialization of this array; all of the members that have the first dimension set to 4 or greater are set to 0 by default. We use two nested *for* loops to print out every member of this two-dimensional array with the proper spacing.

Initializing a two-dimensional character array such as *ch2* is a little easier. We have all of the initializers contained within a set of curly brackets. Within these brackets, there are three strings, one for each of the three first dimension values of the array. Each string is enclosed in a set of double quotation marks, which is in turn enclosed within a set of curly brackets. These are separated by commas. Observe that when we use the *printf* statement to display these three strings, we define the strings as *ch[0]*, *ch[1]*, and *ch[2]*. This causes the *printf* statement to begin reading at the addresses of *ch2[0][0]*, *ch2[1][0]*, and *ch2[2][0]*, respectively, and in each case read until the string terminator is encountered.

7

Using Loops

One of the most frequently encountered situations in programming occurs when you want to repeat a section of the program a number of times, with some changes in conditions at each iteration. The section of code to be repeated is called a loop. The C language includes three powerful methods for controlling program looping: the *while* loop, the *do-while* loop, and the *for* loop.

The *while* Loop

Here is a simple little program making use of a *while* loop:

```
#include <stdio.h>
#include <math.h>

void main(void)
{
      int i=0;

      while (i < 10)
      {
            printf("\n%d   %d", i, i*i)
            i++;
      }
}
```

The loop iterates 10 times. At each iteration it displays *i* and i^2 on the screen. When you are using the *while* loop, the program iterates (loops) for as long as the condition within the parentheses following the *while* statement is true. It is important to note that you as programmer have two responsibilities. First, you must assure that the loop begins with the condition true. If this isn't the case, the program will never enter the loop, not even once. In the sample program, this is taken care of by initializing *i* to 0. Second, you must assure that at some point the condition will become false. If this doesn't occur, your program will remain in the loop forever and will never be able to do anything else useful. In the sample program, this is taken care of by incrementing the value of *i* at every pass through the loop. One of the common mistakes of beginning programmers, when using a *while* loop, is to fail to change the variable that is used to satisfy the condition for exiting the loop, resulting in infinite looping. The mechanics of the *while* loop is very simple. Upon encountering the *while* statement, the program makes the test defined by the set of parentheses following the statement. If the condition is true, the program executes all of the statements between the curly brackets following the condition. (If there is only one statement to be executed, you may omit the curly brackets altogether.) It then returns to the point just following the *while* statement and runs the test again. As long as the condition is true, the section of code in the curly brackets is iterated. When the condition finally becomes false, the program leaves the loop and proceeds to the next statement in the code.

The test that follows the *while* statement can be as complex as you want to make it. The only things that you have to make sure of are that the condition is true initially to permit entering the loop for the first time and that at some point during iteration the condition becomes false so that you can leave the loop.

The Mandelbrot Set

Now we are going to try a more complicated and useful application of the *while* loop. Figure 7-1 lists a program for drawing the Mandelbrot set fractal using several instances of the *while* loop. The Mandelbrot set is probably the most famous fractal curve. It was discovered by the mathematician Benoit Mandelbrot. The picture of the Mandelbrot set is drawn in the following way. Each pixel on the screen represents the location of a particular complex number *c* where the left edge of the screen is the minimum value of the real part of c (*Pmin*), the right-hand edge of the screen is the maximum value of the real part of c (*Pmax*), the bottom edge of the screen is the minimum value of the imaginary part of c (*Qmin*) and the top edge of the screen is the maximum value of the imaginary part of c (*Qmax*). For each value of c we iterate the equation

$$z_{n+1} = z_n^2 + c$$

(Equation 7-1)

where z is also a complex number (starting with $z_0=0$) until the sum of the squares of the real and imaginary parts of z is greater than or equal to 4. At this point, we stop and color the pixel represented by the current value of c in accordance with the number of iterations that have taken place.

The first thing to note is that in order to achieve the most efficiency, we need to precompute all possible values of Q and store them in an array. This is our first application of the *while* loop. Note that as part of the test, we first increment the value of *row* and then test whether it is less than the value of *yres*, the y resolution. Before entering the loop, *row* is set to -1, so that the first iteration occurs with *row* at 0.

The main part of the program consists of three nested *while* loops. The outermost loop iterates once for each column of the display screen. Before this loop begins, the column designator *col* is set to 0. When *col* reaches *xres* we have reached the right-hand edge of the screen and the loop terminates. The loop begins by setting up the proper value of P, the real part of the complex number c in Equation 7-1. At the end of the loop, *col* is incremented in preparation for the next iteration. The next *while* loop iterates once for each row on the screen. Just before entering this loop, the row designator *row* is set to 0. At the end of each iteration of this loop, *row* is incremented in preparation for the next iteration. When *row* reaches *yres* we have reached the bottom of the screen and the loop terminates. Between these two loops, we have covered every pixel location that exists on the display screen.

The innermost *while* loop does all the work of solving the iterated equation. Just before entering this loop, the parameters *x, y, old_x, old_y,* and *color* are all initialized to 0. The test condition for this loop is a little complicated. For the code within the loop to be run, two things must occur. First, the iteration number (value of *color*) must be less than 64. Second, the value of the sum of the squares of x and y (the real and imaginary parts of z) must be less than 4. When either of these conditions becomes false, the loop terminates. Mathematically, if the second condition becomes false, we know that continued iteration of the loop is going to cause the value to blow up (go to infinity). If the first condition becomes false, we know that we have completed the number of iterations that is specified as maximum and the equation still hasn't blown up. It is either growing very very slowly or has reached some sort of stable condition. In either case, we don't want to continue iterating. Within this loop is code that checks for a repeated pattern of values. If this occurs at any point, we know the equation won't blow up and set the number of iterations to maximum so that the loop will terminate on the next pass.

In this program, you have seen how multiple *while* loops can be used to perform a fairly complex programming application. The program generates the traditional Mandelbrot set pattern in Plate 1. You can change the values of *Pmax, Pmin, Qmax*, and *Qmin* to expand or contract this pattern or to select another part of the cosine fractal display. If you don't have a math coprocessor in your computer, the program will run quite slowly, but the result is worth waiting for and watching the actual creation of the display is quite interesting.

Figure 7-1. Program for Generating Mandelbrot Set

```
/*

        MANDLBRO = Program to generate Mandelbrot set

             By Roger T. Stevens   8-21-91

*/

#include <dos.h>
#include <stdio.h>
#include <math.h>

int xres = 640, yres = 480;
```

(continued)

```
int  color, row, col;
float Pmax=1.6, Pmin= -2.2, Qmax = 1.6, Qmin = -1.6;
float Q[480], P, deltaP, deltaQ, x, y, old_x, old_y, xsq,
      ysq;

void plot(int x,int y,int color);
void setmode(int mode);
void cls(int color);

void main(void)
{
      setmode(18);
      cls(7);
      deltaP = (Pmax - Pmin)/xres;
      deltaQ = (Qmax - Qmin)/yres;
      row = 0;
      while (row<yres)
      {
            Q[row] = Qmin + row*deltaQ;
            row++;
      }
      col = 0;
      while (col<xres)
      {
            P = Pmin + col*deltaP;
            row = 0;
            while (row<yres)
            {
                  x = y = old_x = old_y = 0.0;
                  color = 0;
                  while (color<64)
                  {
                        xsq = x*x;
                        ysq = y*y;
                        if (xsq + ysq > 4)
                              break;
                        y = 2*x*y + Q[row];
                        x = xsq - ysq + P;
                        if ((x == old_x) && (y == old_y))
                        {
                              color = 0;
```

(continued)

```
                                        break;
                                }
                                if ((color % 8) == 0)
                                {
                                        old_x = x;
                                        old_y = y;
                                }
                                color ++;
                        }
                        plot(col, row, (color % 16));
                        row++;
                }
                col++;
        }
        getch();
}

/*
╔══════════════════════════════════════════════╗
║                                              ║
║          setmode() = Sets video mode         ║
║                                              ║
╚══════════════════════════════════════════════╝
*/

void setmode(int mode)
{

        union REGS reg;

        reg.x.ax = mode;
        int86 (0x10,&reg,&reg);
}

/*
╔══════════════════════════════════════════════╗
║                                              ║
║          cls() = Clears the screen           ║
║                                              ║
╚══════════════════════════════════════════════╝
*/

void cls(int color)
```

(continued)

```
{
        union REGS reg;

        reg.x.ax = 0x0600;
        reg.x.cx = 0;
        reg.x.dx = 0x1E4F;
        reg.h.bh = color;
        int86(0x10,&reg,&reg);
}

/*
    plot() = Plots a point on the screen at a designated
     position using a selected color for 16 color modes.
 */

void plot(int x, int y, int color)
{
      #define graph_out(index,val)  {outp(0x3CE,index);\
                                  outp(0x3CF,val);}

      int dummy,mask;
      char far * address;

      address = (char far *) 0xA0000000L + (long)y *
            xres/8L + ((long)x / 8L);
      mask = 0x80 >> (x % 8);
      graph_out(8,mask);
      graph_out(5,2);
      dummy = *address;
      *address = color;
      graph_out(5,0);
      graph_out(8,0xFF);
}
```

The *do-while* loop

The *do-while* loop is very similar to the *while* loop except that the test occurs at the end of the loop instead of at the beginning. Here, modified for use with a *do-while* loop is the same simple program that was used of a *while* loop.

```
#include <stdio.h>
#include <math.h>

void main(void)
{
      int i=0;

      do
      {
            printf("\n%d   %d", i, i*i)
            i++;
      }
      while (i < 10);
}
```

You'll note that this is just a bit more complicated than the previous example, but does exactly the same thing. Why, then, would we ever want to use a *do-while* loop? The answer is that there are certain situations where we arrive at the loop without knowing exactly what value may have been established for the variable or variables used in the test. However, whatever their values may be, we want to make at least one pass through the loop. The *do-while* loop satisfies this requirement, since the code in the loop is run first and then the first test is made. If the condition is false to start with, the loop is not repeated, but it has been run once.

The Hyperbolic Cosine Fractal

We're now going to try a more complicated and useful application of the *do-while* loop. Figure 7-2 lists a program using multiple *do-while* loops to create a fractal display of the hyperbolic cosine. The program shows you how *do-while* loops are used, but it is not very good programming practice, since there is no real requirement to run each loop at least once, so each of the *do-while* loops could be replaced by a simpler *while* loop. While you are examining this program, remember that you should never use a *do-while* loop when only a simple *while* loop is needed and practice simplifying this program in your mind. The picture of the hyperbolic cosine fractal is drawn in the following way. Each pixel on the screen represents the location of a particular complex number *c* where the left edge of the screen is the minimum value of the real part of *c* (*Pmin*), the right-hand edge of the screen is the maximum value of the real part of *c* (*Pmax*), the bottom edge of the screen is the minimum value of the imaginary part of *c* (*Qmin*) and the top edge of the screen is the maximum value of the imaginary part of *c* (*Qmax*). For each value of *c* we iterate the equation

$$z_{n+1} = \cosh(z_n) + c$$

(Equation 7-2)

where *z* is also a complex number (starting with $z_0=0$) until the sum of the squares of the real and imaginary parts of *z* is greater than or equal to 100. At this point, we stop and color the pixel represented by the current value of *c* in accordance with the number of iterations that have taken place. If you have a compiler whose library includes hyperbolic functions of complex angles you can simplify the math of this program a lot. The program, as written, makes use of the trigonometric identity

$$\cosh(x + iy) = \cosh x \cos y - i \sinh x \sin y$$

(Equation 7-3)

which permits you to use ordinary hyperbolic functions available with any C compiler.

The first thing to note is that in order to achieve the most efficiency, we need to precompute all possible values of *Q* and store them in an array. This is our first application of the *do-while* loop. We start with *row* set to 0. At the end of each iteration of the loop, we increment *row* then test whether it is less than the value of *yres*, the *y* resolution. When *row* reaches this value, the loop terminates.

The main part of the program consists of three nested *do-while* loops. The outermost loop iterates once for each column of the display screen. Before this loop begins, the column designator *col* is set to 0. When *col* reaches *xres* we have reached the right-hand edge of the screen and the loop terminates. The loop begins by setting up the proper value of *P*, the real part of the complex number *c* in Equation 7-1. At the end of the loop, *col* is incremented in preparation for the next iteration. The next *do-while* loop iterates once for each row on the screen. Just before entering this loop, the row designator *row* is set to 0. At the end of each iteration of this loop, *row* is incremented in preparation for the next iteration. When *row* reaches *yres* we have reached the bottom of the screen and the loop terminates. Between these two loops, we have covered every pixel location that exists on the display screen.

The innermost *do-while* loop does all the work of solving the iterated equation. Just before entering this loop, the parameters *x, y, old_x, old_y,* and *color* are all initialized to 0. The test condition for this loop is a little complicated. For the code within the loop to be run, two things must occur. First, the iteration number (value of *color*) must be less than 64. Second, the value of the sum of the squares of x and y (the real and imaginary parts of z) must be less than 4. When either of these conditions becomes false, the loop terminates. Mathematically, if the second condition becomes false, we know that continued iteration of the loop is going to cause the value to blow up (go to infinity). If the first condition becomes false, we know that we have completed the number of iterations that is specified as maximum and the equation still hasn't blown up. It is either growing very very slowly or has reached some sort of stable condition. In either case, we don't want to continue iterating. Within this loop is code that checks for a repeated pattern of values. If this occurs at any point, we know the equation won't blow up and set the number of iterations to maximum so that the loop will terminate on the next pass.

In this program, you have seen how multiple *do-while* loops can be used to perform a fairly complex programming application. Remember, however, when you are programming do not use the *do-while* loop if a simpler *while* loop will suffice. The hyperbolic cosine fractal program generates the interesting picture shown in Plate 2. You can change the values of *Pmax, Pmin, Qmax,* and *Qmin* to expand or contract this pattern or to select another part of the hyperbolic cosine fractal display. If you don't have a math coprocessor in your computer, have patience, because the program will run quite slowly.

Figure 7-2. Using the do-while Loop to Generate the Hyperbolic Cosine Fractal

```
/*
╔═══════════════════════════════════════════════════════════════╗
║                                                               ║
║      COSHFRAC = Program to generate hyperbolic cosine         ║
║                       fractal set                             ║
║                                                               ║
║              By Roger T. Stevens  5-29-92                     ║
║                                                               ║
╚═══════════════════════════════════════════════════════════════╝
*/
#include <dos.h>
#include <stdio.h>
#include <math.h>

int xres = 640, yres = 480;
int  color, row, col;
float Pmax=2.6, Pmin= 1.9, Qmax = -1.15, Qmin = -1.8;
float Q[480], P, deltaP, deltaQ, old_x, old_y, temp, x, y,
      xsq, ysq;

void plot(int x,int y,int color);
void setmode(int mode);
void cls(int color);

main()
{
      setmode(16);
      cls(7);
      deltaP = (Pmax - Pmin)/xres;
      deltaQ = (Qmax - Qmin)/yres;
      row = 0;
      do
            Q[row] = Qmax - row*deltaQ;
      while (++row<yres);
      col = 0;
      do
      {
            P = Pmin + col*deltaP;
            row = 0;
```

(continued)

```
                do
                {
                        x = y = old_x = old_y = 0.0;
                        color = 0;
                        do
                        {
                                temp = cos(x)*cosh(y) + P;
                                y = -sin(x)*sinh(y) + Q[row];
                                x = temp;
                                if ((x == old_x) && (y == old_y))
                                {
                                        color = 0;
                                        break;
                                }
                                if ((color % 8) == 0)
                                {
                                        old_x = x;
                                        old_y = y;
                                }
                        }
                        while ((++color<64) && ((x*x + y*y) <
                                100));
                        plot(col, row, (color % 16));
                        row++;
                }
                while (row<yres);
                col++;
        }
        while (col<xres);
        getch();
}

/*
 ╔══════════════════════════════════════════════════════╗
 ║                                                      ║
 ║            setmode() = Sets video mode               ║
 ║                                                      ║
 ╚══════════════════════════════════════════════════════╝
*/

void setmode(int mode)
{
```

(continued)

```
    union REGS reg;

    reg.x.ax = mode;
    int86 (0x10,&reg,&reg);
}

/*
╔══════════════════════════════════════════════════════════╗
║                                                          ║
║              cls() = Clears the screen                   ║
║                                                          ║
╚══════════════════════════════════════════════════════════╝
*/

void cls(int color)
{
    union REGS reg;

    reg.x.ax = 0x0600;
    reg.x.cx = 0;
    reg.x.dx = 0x1E4F;
    reg.h.bh = color;
    int86(0x10,&reg,&reg);
}

/*
╔══════════════════════════════════════════════════════════╗
║                                                          ║
║  plot() = Plots a point on the screen at a designated    ║
║   position using a selected color for 16 color modes.    ║
║                                                          ║
╚══════════════════════════════════════════════════════════╝
*/

void plot(int x, int y, int color)
{
    #define graph_out(index,val)  {outp(0x3CE,index);\
                                   outp(0x3CF,val);}

    int dummy,mask;
    char far * address;

    address = (char far *) 0xA0000000L + (long)y *
```

(continued)

```
                xres/8L + ((long)x / 8L);
        mask = 0x80 >> (x % 8);
        graph_out(8,mask);
        graph_out(5,2);
        dummy = *address;
        *address = color;
        graph_out(5,0);
        graph_out(8,0xFF);
}
```

The *for* Loop

The *for* loop is the most sophisticated of the loop constructions. It can usually make your loop programming simpler and faster. The *for* command consists of the word *for* followed by a set of parentheses containing three separate sections. The first section consists of initializers. In this section, initial values can be established for any parameters that are used within the loop. As many parameters as desired may be initialized. They are separated by commas. The second section consists of the test that determines whether the loop should continue to be iterated. This may be as complex a test as you can devise and test as many variables as desired, but it must be defined as a single test. The third section modifies parameters that are to be changed at each pass through the loop. As many parameters as desired may be modified, with each modification expression separated by commas. The three sections of the *for* loop are separated by semi-colons. Here's the same simple program that we used for the other two loops, adapted to the *for* loop.

```
#include <stdio.h>
#include <math.h>

void main(void)
{
        int i;

        for (i=0; i<10; i++)
                printf("\n%d   %d", i, i*i)
}
```

In this example you can clearly see how the *for* loop is used in a simple situation. In the initialization section the parameter i is set to 0 at the beginning of the loop. In the test section, i is tested to see whether it is less than 10 and as long as this is so, the loop continues to iterate. When i becomes equal to or greater than 10, the loop terminates. In the modification section, i is incremented at each pass through the loop.

The Legendre Polynomial Fractal

Next, we're going to look at a simple program to generate a Legendre Polynomial fractal set using some of the sophistication of *for* loops. This program is listed in Figure 7-3. The Legendre Polynomials are a set of polynomials that occur in advanced calculus applications. You can find a listing of these polynomials in many standard mathematical reference books. Each one can be used in an iterated equation to produce a unique fractal set. The Legendre polynomial that we're going to use in this example is P_3, which is defined as

$$P_3 = \frac{5z^3 - 3z}{2}$$

(Equation 7-4)

The corresponding iterated equation for fractal generation is

$$z_{n+1} = \frac{5z_n^3 - 3z_n}{2} + c$$

(Equation 7-5)

The program is structured in a very similar way to the two fractal programs already described in this chapter. You won't notice very much difference between the two outermost *for* loops and the corresponding *while* and *do-while* loops of the previous programs. It is in the innermost loop where the significant differences occur. The definition of the *for* loop includes much of the code that was previously included elsewhere in the program. Let's look in detail at what this loop is telling us. The first section tells us that at the beginning of the loop, we initialize *color, x, y, old_x, old_y, xsq* and *ysq* to 0. The second section indicates that we test for both *color* to be less than 64 and *xsq* + *ysq* to be less than 4.0. The third section says that at the beginning of each pass through the loop we increment *color*, set *xsq* to *x*x*, and set *ysq* to *y*y*. Note that having all of this information right there in the definition of the *for* loop gives us a very clear picture of exactly how the loop works in this situation. It also makes the code more compact and may actually speed up operation. This illustrates how you can do a better job of programming by using *for* loops in many situations. The picture resulting from running the Legendre program is shown in Plate 3.

Figure 7-3. Using a for Loop to Generate the Legendre Polynomial Fractal

```
/*

    LEGENDRE = Program to generate cosine Legendre P3
                        fractal set.

                By Roger T. Stevens  6-14-92

*/

#include <dos.h>
#include <stdio.h>
#include <math.h>

int xres = 640, yres = 480;
int  color, row, col;
float Pmax = 1.0, Pmin = -1.0, Qmax = 0.4, Qmin = -0.4;
float Q[480], P, deltaP, deltaQ, old_x, old_y, temp, x, y,
      xsq, ysq;

void plot(int x,int y,int color);
void setmode(int mode);
void cls(int color);

main()
{
      setmode(18);
      cls(7);
      deltaP = (Pmax - Pmin)/xres;
```

(continued)

```
        deltaQ = (Qmax - Qmin)/yres;
        for (row=0; row<yres; row++)
                Q[row] = Qmax - row*deltaQ;
        for (col=0; col<xres; col++)
        {
                P = Pmin + col*deltaP;
                for (row=0; row<yres; row++)
                {
                        x = y = old_x = old_y = 0.0;
                        for (color=0; ((color<64) && ((x*x +
                                y*y) < 100)); color++)
                        {
                                temp = 0.5*(5.0*x*x*x - 15.0*x*y*y
                                        - 3.0*x) + P;
                                y = 0.5*(15.0*x*x*y - 5.0*y*y*y) +
                                        Q[row];
                                x = temp;
                                if ((x == old_x) && (y == old_y))
                                {
                                        color = 0;
                                        break;
                                }
                                if ((color % 8) == 0)
                                {
                                        old_x = x;
                                        old_y = y;
                                }
                        }
                        plot(col, row, (color % 16));
                }
        }
        getch();
}

/*
+--------------------------------------------------------+
|                                                        |
|           setmode() = Sets video mode                  |
|                                                        |
+--------------------------------------------------------+
*/
```

(continued)

```
void setmode(int mode)
{
      union REGS reg;

      reg.x.ax = mode;
      int86 (0x10,&reg,&reg);
}

/*
            cls() = Clears the screen
*/

void cls(int color)
{
      union REGS reg;

      reg.x.ax = 0x0600;
      reg.x.cx = 0;
      reg.x.dx = 0x1E4F;
      reg.h.bh = color;
      int86(0x10,&reg,&reg);
}

/*
  plot() = Plots a point on the screen at a designated
  position using a selected color for 16 color modes.
*/

void plot(int x, int y, int color)
{
      #define graph_out(index,val)  {outp(0x3CE,index);\
                                     outp(0x3CF,val);}

      int dummy,mask;
      char far * address;
```

(continued)

```
        address = (char far *) 0xA0000000L + (long)y *
              xres/8L + ((long)x / 8L);
        mask = 0x80 >> (x % 8);
        graph_out(8,mask);
        graph_out(5,2);
        dummy = *address;
        *address = color;
        graph_out(5,0);
        graph_out(8,0xFF);
}
```

8

Conditional Statements

One of the key features of any programming language is the capability to set up a test condition, which, if true, causes the program to perform one action or set of actions and, if false, causes an entirely different action or set of actions to be performed. Programs that can't make choices are pretty uninteresting; programs that can analyze a situation and take different actions, depending on the conditions that exist, begin to approach the ability of human beings to analyze, make decisions, and take actions thereon. In addition to making tests for the running of loops, which is covered in Chapter 7, C has three ways in which the decision-making process can take place. The first is the *if-else* statement. The second uses the punctuation marks *?* and *:*. The third is the *switch-case* statement, which is used when different actions are taken for a lot of different values of one integer or character type argument.

The *if-else* Statement

The *if-else* statement begins with the word *if* followed by a test contained within parentheses. (Note that the close parenthesis is **not** followed by a semicolon.) Following this test is a statement, or series of statements, that are to be performed only if the condition of the test is true. If only a single statement is to be performed when the condition is true, it only needs to be listed following the condition; if more than one statement is involved, then all statements to be performed when the condition is true should be included in curly brackets. Here is an example of a simple *if* statement followed by a single action

```
if (r > 4)
        printf("R is larger than 4.");
```

When more than one statement is involved, the *if* statement looks like this

```
if (r  > 4)
{
        printf("R is larger than 4.");
        r++;
}
```

The *else* part of the *if-else* statement is optional. Thus the two examples just given stand by themselves as perfectly correct and acceptable usage of the *if* statement. If, after one of these constructions the word *else* appears, then a single statement (or multiple statements within curly brackets) that follow the *else* will be executed only if the condition of the *if* statement is false. For example

```
if (r > 4)
{
        printf("R is greater than 4.");
        r--;
}
else
{
        printf("R is less than or equal to 4.");
        r++;
}
```

This little section of code displays "R is greater than 4." and decrements *r* if *r* is greater than 4; if *r* is 4 or less, the alternate action takes place, "R is less than or equal to 4." is displayed and *r* is incremented. You're free to nest as many *if-else* statements as necessary to make whatever decision you are interested in, but you need to make sure that you use whatever sets of curly brackets are needed so that there is no ambiguity as to which *if* statement a particular *else* is associated with.

Relational Operators

You've already had experience with one relational operator in the examples just given. The operator > says that you test the value on the left to see if it is *greater than* the value on the right. The following is a list of the relational operators available in C

Operator	Meaning
==	equal to
!=	not equal to
>	greater than
>=	greater than or equal to
<	less than
<=	less than or equal to

You need to be especially careful to note that the relational *equal to* operator is written with a double equals sign (==). A common error is to write a piece of code like the following

```
if (a = 4)
        printf("a is 4.");
```

What you intended to do was test to see whether *a* is equal to 4 and if so to display the text. However, since you didn't use the double equals sign, you are actually assigning *a* the value of 4, so that the text will always be printed, and no matter what value *a* had when you entered this section of code, it will always be 4 when you are finished. If this section of code is within a loop where you are planning to increment *a* at every pass and exit when *a* reaches some predetermined value, you are in especially bad trouble because *a* is always going to be reset to 4, so you will never exit from the loop.

More Complicated *if* Statements

You can combine as many tests as you want into a single complex test within an *if* statement by using either of two logical operators: ||

means that two tests are ORed together; *&&* means that two tests are ANDed together. Thus the *if* statement

```
if ((a==3) || (b < 4)) &&((c <= 2) || (d >= 9))
      r = 7;
```

means that the test is true if *a* is equal to 3 or *b* is less than 4 and at the same time *c* is less than or equal to 2 or *d* is greater than or equal to 9. There is no limit as to how complicated you can make the test in an *if* statement; just don't get carried away and make the conditions so complicated that the overall test can never be true or never be false.

The Julia Set

If we take the same iterated equation given in Equation 7-1 and instead of varying *c* we hold *c* constant and instead of starting each set of iterations with $z_0=0$ we vary *z* over a range in the complex plane, we obtain a totally different fractal picture known as the Julia set. Figure 8-1 is a listing of a program to generate a Julia set, making use of *if-else* statements together with *goto* statements to perform all of the necessary looping. This isn't very good programming practice; normally you should use one of the looping techniques described in the previous chapter and avoid any *goto* statements whatsoever. However, it does give you a good demonstration of how *if-else* statements can be used. After initialization, the program begins the display process by comparing *col* with *xres* and performing everything within the curly brackets if *col* is less than *xres*. The very last statements increment *col* and return to the label *cols*, which is just before the *if* statement. Thus this section of code repeats until *col* has increased to *xres* at which point the code within the curly brackets is skipped, so another iteration doesn't occur. Similarly, the next *if* statement causes a looping that iterates for each value of *row* from 0 to *yres*. The code following the next *if* statement is executed if both *color* is less than 64 and the sum of *xsq* and *ysq* is less than 4. At the close of this code, *color* is incremented and the program jumps back to the label *iterations*, which is just before the *if* statement. Thus this whole section repeats until either the second part of the test becomes false (which is due to the iterated function starting to blow up) or until 64 iterations have taken place (at which time the first part of the test becomes false).

You'll find that there are lots of interesting Julia sets, which you can investigate by changing the values for *P* and *Q*. You can change the magnification of the particular Julia set that you are viewing by changing the values of *Xmax, Xmin, Ymax,* and *Ymin*. The one defined by the current program values is shown in Plate 4.

Figure 8-1. Program to Generate Julia Set

```
/*
╔══════════════════════════════════════════════════════════╗
║                                                          ║
║         JULIA = Program to generate Julia set            ║
║                                                          ║
║            By Roger T. Stevens  6-10-92                  ║
║                                                          ║
╚══════════════════════════════════════════════════════════╝
*/

#include <dos.h>
#include <stdio.h>
#include <math.h>

int xres = 640, yres = 480;
int  color, row, col;
float Xmax=1.8, Xmin= -1.8, Ymax = 1.2, Ymin = -1.2;
float P=-0.053750, Q=0.653000, deltaX, deltaY, x, y, xsq,
      ysq;

void plot(int x,int y,int color);
void setmode(int mode);
void cls(int color);

main()
{
      setmode(18);
      cls(7);
      deltaX = (Xmax - Xmin)/xres;
      deltaY = (Ymax - Ymin)/yres;
      col = 0;
      cols:
      if (col<xres)
```

(continued)

```
        {
                row = 0;
                rows:
                if (row<yres)
                {
                        x = Xmin + col*deltaX;
                        y = Ymax - row*deltaY;
                        xsq = ysq = 0.0;
                        color = 0;
                        iterations:
                        if ((color<64) && (xsq + ysq < 4))
                        {
                                xsq = x*x;
                                ysq = y*y;
                                y = 2*x*y + Q;
                                x = xsq - ysq + P;
                                color++;
                                goto iterations;
                        }
                        if (color >= 64)
                                color = ((int) ((xsq + ysq) *
                                        8.0)) %7 + 9;
                        else
                                color = 1;
                        plot(col, row, color);
                        row++;
                        goto rows;
                }
        col++;
        goto cols;
        }
        getch();
}

/*
┌──────────────────────────────────────────────┐
│                                              │
│          setmode() = Sets video mode         │
│                                              │
└──────────────────────────────────────────────┘
*/
```

(continued)

```
void setmode(int mode)
{
      union REGS reg;

      reg.x.ax = mode;
      int86 (0x10,&reg,&reg);
}

/*
                  cls() = Clears the screen
*/

void cls(int color)
{
       union REGS reg;

       reg.x.ax = 0x0600;
       reg.x.cx = 0;
       reg.x.dx = 0x1E4F;
       reg.h.bh = color;
       int86(0x10,&reg,&reg);
}

/*
   plot() = Plots a point on the screen at a designated
     position in a selected color for 16 color modes.
*/

void plot(int x, int y, int color)
{
      #define graph_out(index,val)  {outp(0x3CE,index);\
                                     outp(0x3CF,val);}

      int dummy,mask;
      char far * address;
```

(continued)

```
        address = (char far *) 0xA0000000L + (long)y *
              xres/8L + ((long)x / 8L);
        mask = 0x80 >> (x % 8);
        graph_out(8,mask);
        graph_out(5,2);
        dummy = *address;
        *address = color;
        graph_out(5,0);
        graph_out(8,0xFF);
}
```

The ? : Trinary Conditional Function and the Dragon Curve

There is one special kind of conditional statement that can be represented in a shorthand way that makes for compact and hopefully easier to read code. If such a statement were written using the *if-else* method just described, it would look like this

```
if (a < 4)
      d = 6;
else
      d = 3;
```

They key here is that a single variable is set to one of two different values, depending upon whether the result of a test is true or false. You can get much more complex than this, but you are limited to alternative values for a single variable. The new technique obtains the same result like this

```
d = (a < 4) ? 6 : 3;
```

The C language automatically understands that if it sees a *?* it should consider what's within the parentheses preceding it as a test. If the result of the test is true, the variable on the left of the equals sign is assigned the value following the *?* (which may be a complicated equation); if the test result is false, the variable to the left of the equals sign is assigned the value following the *:*. As long as you don't get so complex as to be impossible to understand, this technique provides a compact and elegant way of replacing these special cases of the *if-else* statement. (Note, however, that there are lots of things that can be done with *if-else* statements that can't be done using this technique.)

The dragon curve is very similar to the Julia set, except that it is obtained by iterating the equation

$$z_{n+1} = cz_n(1-z_n) \qquad \text{(Equation 8-1)}$$

The program to generate a dragon curve is listed in Figure 8-2. You'll observe that it is very similar to the Julia set program except for the mathematics and for the fact that the color to be displayed is obtained by a *? :* statement instead of an *if-else* statement. Note that for this situation, the technique used to obtain the color in the dragon set program is a distinct improvement. The result of running this program is shown in Plate 5. You can make some very different dragon curves by making very small changes in the values of *P* and *Q*.

Figure 8-2. Program to Generate a Dragon Curve

```
/*

        DRAGON = Program to generate a dragon curve

              By Roger T. Stevens   6-10-92

*/

#include <dos.h>
#include <stdio.h>
#include <math.h>

int xres = 640, yres = 480;
int  color, row, col;
float Xmax=1.4, Xmin= -0.4, Ymax = 0.8, Ymin = -0.8;
float P=1.646, Q=0.967, deltaX, deltaY, sqdif, prod, x, y,
      y2, xsq, ysq;

void plot(int x,int y,int color);
void setmode(int mode);
```

(continued)

```
void cls(int color);
main()
{
      setmode(18);
      cls(7);
      deltaX = (Xmax - Xmin)/xres;
      deltaY = (Ymax - Ymin)/yres;
      col = 0;
      cols:
      if (col<xres)
      {
            row = 0;
            rows:
            if (row<yres)
            {
                  x = Xmin + col*deltaX;
                  y = Ymax - row*deltaY;
                  xsq = ysq = 0.0;
                  color = 0;
                  iterations:
                  if ((color<256) && (xsq + ysq < 4))
                  {
                        xsq = x*x;
                        ysq = y*y;
                        sqdif = ysq - xsq;
                        prod = 2*x*y;
                        y2 = Q*(sqdif + x) - P*(prod - y);
                        x = P*(sqdif + x) + Q*(prod - y);
                        y = y2;
                        color++;
                        goto iterations;
                  }
                  color = (color >= 256) ? ((int) ((xsq +
                        ysq) * 8.0)) %7 + 9 : 1;
                  plot(col, row, color);
                  row++;
                  goto rows;
            }
      col++;
      goto cols;
      }
```

(continued)

```
    getch();
}

/*
┌──────────────────────────────────────────────────┐
│                                                  │
│          setmode() = Sets video mode             │
│                                                  │
└──────────────────────────────────────────────────┘
*/

void setmode(int mode)
{
    union REGS reg;

    reg.x.ax = mode;
    int86 (0x10,&reg,&reg);
}

/*
┌──────────────────────────────────────────────────┐
│                                                  │
│          cls() = Clears the screen               │
│                                                  │
└──────────────────────────────────────────────────┘
*/

void cls(int color)
{
    union REGS reg;

    reg.x.ax = 0x0600;
    reg.x.cx = 0;
    reg.x.dx = 0x1E4F;
    reg.h.bh = color;
    int86(0x10,&reg,&reg);
}
```

(continued)

```
/*
    plot() = Plots a point on the screen at a designated
      position in a selected color for 16 color modes.
*/

void plot(int x, int y, int color)
{
      #define graph_out(index,val)  {outp(0x3CE,index);\
                                      outp(0x3CF,val);}

      int dummy,mask;
      char far * address;

      address = (char far *) 0xA0000000L + (long)y *
            xres/8L + ((long)x / 8L);
      mask = 0x80 >> (x % 8);
      graph_out(8,mask);
      graph_out(5,2);
      dummy = *address;
      *address = color;
      graph_out(5,0);
      graph_out(8,0xFF);
}
```

The *switch-case* Statement

The *switch-case* statement is used when one control argument, which must be an integer or character, can take on a lot of different values, each of which determines a different action on the part of the program. Here is a simple example

```
switch(number)
{
      case 2:
            a = 4;
            b = 5;
            break;
      case 3:
            a = 6;
```

```
        case 4:
        case 5:
              b = 7;
              c = 8;
              break;
        default:
              a = 0;
              b = 0;
              c = 0;
}
```

The keyword *switch* is followed by the name of the control variable in parentheses (in this case *number*). Next, each *case* covers what happens when *number* takes that particular value. For a value of 2, *a* is set to 4 and *b* to 5. Then a *break* occurs that causes the program to leave the *switch* section of code. For a value of 3, *a* is set to 6 and since there is no *break* the program continues, setting *b* to 7 and *c* to 8. These last two actions are also performed for *case* values of 4 and 5. Then another *break* occurs, which terminates action for *number* of 3, 4, or 5. Finally, if *number* takes on any other value, the *default* case is run, which sets *a, b*, and *c* to 0. You can have just about as many cases as you want to. The *default* case is optional; if it doesn't appear, the *switch* code does nothing for values that aren't listed.

Figure 8-3 is a listing of the *mandelsw* program which generates an enlarged view of part of the Mandelbrot set. It is similar to the Mandelbrot set program of Chapter 7, except for the bounds and for the use of a maximum of 512 iterations instead of 64. You're going to see this same program again in Chapter 9, where ordinary colors are used and four different enlargements are allowed. How colors are assigned to the Mandelbrot set makes a great deal of difference in the appearance of the picture and what features are emphasized. You can compare the result of running this program, which is shown in Plate 6, with the ordinarily colored version from the next chapter, which is shown in Plate 7. The selection of colors used in this program is a natural for the use of the *switch-case* statement, as you will note in the program listing.

Figure 8-3. Use of switch-case Statement in Mandelbrot Set

```
/*
 ┌─────────────────────────────────────────────────────────┐
 │                                                         │
 │    MANDELSW = Program to generate Mandelbrot set        │
 │                  with unusual colors.                   │
 │                                                         │
 │              By Roger T. Stevens  9-10-92               │
 │                                                         │
 └─────────────────────────────────────────────────────────┘
*/

#include <dos.h>
#include <stdio.h>
#include <math.h>

int xres = 640, yres = 480;
int  color, i, row, col;
double Pmax=-0.745418, Pmin= -0.745438, Qmax = 0.113016,
      Qmin = 0.113000;
double Q[480], P, deltaP, deltaQ, x, y, xsq, ysq;

void plot(int x,int y,int color);
void setmode(int mode);
void cls(int color);

void main()
{
      int i;
      setmode(18);
      cls(0);
      cls(7);
      deltaP = (Pmax - Pmin)/xres;
      deltaQ = (Qmax - Qmin)/yres;
      for (row=0; row<yres; row++)
            Q[row] = Qmin + row*deltaQ;
      for (col=0; col<xres; col++)
      {
            P = Pmin + col*deltaP;
            for (row=0; row<yres; row++)
            {
```

(continued)

```
                    x = y = xsq = ysq = 0.0;
                    for (i=0; ((i<512) && (xsq + ysq < 4));
                          i++)
                    {
                          xsq = x*x;
                          ysq = y*y;
                          y = 2*x*y + Q[row];
                          x = xsq - ysq + P;
                    }
                    color = i / 100;
                    switch(color)
                    {
                          case 0:
                                color = 2;
                                break;
                          case 1:
                                color = 9;
                                break;
                          case 2:
                                color = 2;
                                break;
                          case 3:
                                if (i % 2)
                                      color = 13;
                                else
                                      color = 19;
                                break;
                          case 4:
                                color = 14;
                                break;
                          case 5:
                                color = 0;
                    }
                    plot(col, row, (color % 16));
              }
        }
        getch();
  }
```

(continued)

```
/*
┌──────────────────────────────────────────────────────────┐
│                                                          │
│              setmode() = Sets video mode                 │
│                                                          │
└──────────────────────────────────────────────────────────┘
*/

void setmode(int mode)
{
      union REGS reg;

      reg.x.ax = mode;
      int86 (0x10,&reg,&reg);
}

/*
┌──────────────────────────────────────────────────────────┐
│                                                          │
│              cls() = Clears the screen                   │
│                                                          │
└──────────────────────────────────────────────────────────┘
*/

void cls(int color)
{
       union REGS reg;

       reg.x.ax = 0x0600;
       reg.x.cx = 0;
       reg.x.dx = 0x1E4F;
       reg.h.bh = color;
       int86(0x10,&reg,&reg);
}

/*
┌──────────────────────────────────────────────────────────┐
│                                                          │
│  plot() = Plots a point on the screen at a designated    │
│    position in a selected color for 16 color modes.     │
│                                                          │
└──────────────────────────────────────────────────────────┘
*/
```

(continued)

```
void plot(int x, int y, int color)
{
      #define graph_out(index,val)  {outp(0x3CE,index);\
                                     outp(0x3CF,val);}

      int dummy,mask;
      char far * address;

      address = (char far *) 0xA0000000L + (long)y * xres/8L
            + ((long)x / 8L);
      mask = 0x80 >> (x % 8);
      graph_out(8,mask);
      graph_out(5,2);
      dummy = *address;
      *address = color;
      graph_out(5,0);
      graph_out(8,0xFF);
}
```

9

Output to Screen and Printer

A computer program doesn't do you very much good unless it can produce output, either to a screen, a printer or a disk. The standard C libraries furnished with every C compiler have functions for producing the necessary outputs. In this chapter, we shall learn how to use them.

The *putch* Function

The first function that we shall consider is *putch*. This function outputs a character to the screen at the location that is currently stored for the cursor. (If you are in one of the text modes, you will see the cursor on the screen and this is where the character from *putch* will appear. If you are in a graphics mode, the cursor doesn't appear on the screen, but its location is stored nevertheless and indicates where the character appears on the screen.) The actual code line will be written like this

```
putch(ch);
```

where *ch* is of type *int*. The fact that the type of the passed parameter is *int* (an integer) is somewhat misleading, since actually only the four least significant bits of the parameter are used to determine the character to be displayed. Thus you can just pass something of type *char* (a character) and everything will work out fine. You can also define the character to be passed in three different ways:

```
putch('R');
```

where *R* is the character that you want to display, or

```
putch(0x53);
```

where the number following the *0x* is the ASCII designation for the desired character, or

```
putch('\x53');
```

where the number following the *\x* is the ASCII designation for the desired character. The latter two methods are useful when you want to display an IBM special character (such as lines and corners for boxes) that cannot be typed from the keyboard. Appendix A gives the ASCII designations for characters available on the IBM PC and compatible computers. Each time that *putch* is called, it increments the cursor location, so that if you use a series of *putch* function calls, the characters will be displayed in a string on the screen, rather than being superimposed on top of each other.

Depending on what compiler you are using, the *putch* function may return nothing or it may return the character sent when display was successful and EOF (end-of-file character) when the output was not successful. If you want your program to be portable between various computers, it should not depend on a return from this function.

It is important to note that the *putch* function works only in the text display modes. If you are using a graphics display mode, the *putch* function will do nothing; to display a character in these modes, use the *putchar* function described below.

The *putc* and *fputc* Functions

The *putc* function is the same as the *putch* function except that it has a second parameter which is the stream to which the character is output. If you define the stream as *stdout*, which is normally the display screen, you'll get the same result as if you were using *putch*.

The code would be

```
putc(ch,stdio);
```

You can also use this function to output to your printer by

```
putc(ch, stdprn);
```

or to a disk file by

```
putc(ch, file);
```

where you have already opened the file *file* for writing.

There is also a function *fputc*, which is exactly the same as *putc* as far as its operation is concerned. The only difference is that *fputc* is a true function, whereas *putc* is a macro that will be inserted into the compiled code at each appropriate location. Normally, in programming, you won't be much concerned about this difference. It will manifest itself by the fact that using *putc* may result in a slightly faster and longer compiled program than using *fputc*.

The *putchar* and *fputchar* Functions

The *putchar* function passes one parameter as follows

```
putchar(ch);
```

It is the same as *putc* with the stream defined as *stdout*. There is also a corresponding *fputchar* function.

The *puts* and *fputs* Functions

For outputting a whole string of characters in one operation, we have the functions *puts* and *fputs*. These functions are similar to *putc* and *fputc* except that the first argument is the address of a character string, instead of a single character. Furthermore, *puts* only passes a single parameter, the output being fixed to *stdout* (the display). If you have defined a character string as

```
char string[34];
```

and then loaded this array with a string of 33 characters or less (the array must include a NULL character (00) at the end of the string), then you can output it to *stdout* (the display) like this:

```
puts(ch);
```

If you want to include the string within the *puts* statement, you enclose it within double quotation marks rather than the single quotation marks used for the single character function, like this:

```
puts("Welcome to the world of fractals...");
```

If you are using this technique, you can use *x* to include untypeable characters within the string. This statement

```
puts("\x01Welcome to the world of fractals...\x01");
```

will display the sentence on the screen with a smiling face on either side of it. If you want to output a string to the display with *fputs* you would do this

```
fputs(ch,stdout);
```

You can also use this function to output to the printer or a disk file.

The *putw* Function

The *putw* function is similar to *putc* except that it writes a word (of length integer, or 2 bytes) to the specified stream. This enables you to output two characters in one operation. If you have defined an integer by

```
int integer;
```

at the beginning of your program or function, you can then send it to a stream (in this case the display screen) like this

```
putw(integer, stdout);
```

You can also define the two characters to be output within the *putw* statement:

```
putw('HA', stdout);
```

Note that even though there are two characters specified, they are enclosed in single quotation marks as if they were a single character. You can also use *0x* to precede a number representing two ASCII characters:

```
putw(0x4148, stdout);
```

Because of the way in which the PC handles words, the two least significant digits of this 4-digit number will define the first (leftmost) character to be displayed and the two most significant digits will define the last (rightmost) character to be displayed. The *putw* function displays the two characters side by side (not superimposed) and increments the cursor location so that the next character to be displayed will follow the last one written by *putw*.

The *printf* Function

The functions just described are great for outputting characters or strings of characters, but they don't provide much help if we have a variable containing some number and want to display it. The *printf* function gives us a lot of flexibility in formatting and defining numbers and strings and displaying them. This function automatically outputs to *stdout*, which is the display screen unless it has been redirected. The *printf* function can handle a variable number of parameters. The first parameter is always a format string, which is enclosed in double quotation marks. The remaining parameters are variables whose values will be output in the form specified in the format string. You, as programmer, need to make sure that your format string specifies the format of each variable in your parameter list and that you have a variable in the list for each format specifier. If the number of format specifiers and the number of variables do not match, you can get some really strange results from the *printf* statement. Everything in the format string that is not part of a format specifier is output as a character. This enables you to mix fixed data with the current values of variables on your display.

Type Characters

For each variable in the variable list of your *printf* statement, you must have a format specifier in the format string. As a minimum, the format specifier must consist of a % character followed by a *type* character. The *type* character determines the format in which the output is displayed. Table 9-I lists all of the *type* characters and the way in which they display particular arguments. Make sure you use a *type* character that corresponds to the type of argument you desire to process; otherwise the results will be strange and unpredictable. Note that each *type* has a default precision that will be used to display the result if you do not specify otherwise.

The *flag* Field

Several other means are available for further specifying the format in which an output is displayed. The entire format specifier, including all possible options, is of the form

% [flags] [width] [.precision] [size] type

Table 9-II shows the permissible contents of the *flag* field and how each *flag* affects the output display. The *flag* field affects justification, decimal points, trailing zeroes, and prefixes for octal and hexadecimal numbers.

The *width* Field

The *width* field specifies the minimum width of the print zone for a particular *type*. You can use this field to assure that columns of numbers that you are displaying are properly lined up, but you need to remember that this field only specifies the *minimum* width for a print zone. If your parameter value exceeds the width specified, it will not be truncated. Instead, the print zone will be extended making everything that follows on the print line out of whack.

Table 9-1. Type Characters Used with printf Statement

Type	Format	Input Argument
c	Single character.	char or int
d	Signed decimal integer. (+ sign not displayed; - sign displayed.)	int
e	Fixed point number in the form [-]1.567891e+12. The whole number is a 1 or a 0. The default precision of the portion following the decimal point is six decimal places. The exponent (power of 10 to which the preceding is to be raised) follows the *e*.	float or double
E	Fixed point number in the form [-]1.567891E+12. The whole number is a 1 or a 0. The default precision of the portion following the decimal point is six decimal places. The exponent (power of 10 to which the preceding is to be raised) follows the *E*.	float or double
f	Fixed point number in the form [-]1234.567891. The default precision of the portion following the decimal point is 6 decimal places.	float or double
g	Fixed point number is printed in the most compact form, removing all leading and terminating 0's and removing the decimal point if it is not followed by nonzero numbers. Otherwise the form is either *e* or *f*.	float or double
G	Fixed point number is printed in the most compact form, removing all leading and terminating 0's and the decimal point if is not followed by nonzero numbers. Otherwise the form is either *E* or *F*.	float or double

(continued)

Table 9-I. Type Characters Used with printf Statement (cont.)

Type	Format	Input Argument
i	Signed decimal integer. (+ sign not displayed; - sign displayed.)	int
n	Stores at the input argument address the number of characters thus far written out.	int * (address of an integer variable)
o	Unsigned octal integer.	int
p	Displays address of pointer in *X* format. Output will either be *yyyy*, representing offset or *xxxx:yyyy*, where *xxxx* is segment address and *yyyy* is offset, depending on memory model being used and whether an *F* or *N* modifier has been selected.	void* (pointer to any data type)
s	Displays characters until a NULL (00) is encountered or the specified precision is reached.	void * (pointer to any data type)
u	Unsigned decimal integer.	int
x	Unsigned hexadecimal integer using small letters for numbers greater than 9.	int
X	Unsigned hexadecimal integer using small letters for numbers greater than 9.	int
%	Displays the % character.	none

The *precision* Field

The *precision* field, which is optional, consists of a decimal point (period) followed by a number or an asterisk. The asterisk means that the next argument in the string of *printf* arguments (which must be of type *int*) will take the place of the number part of this field. The number has different effects for different types of data to be printed. If the data is of type **d, i, u, o, x**, or **X**, the *precision* field specifies the number of digits to print. If this *precision* is greater than the number of digits in the data value, leading 0's will be printed. The default value for *precision* is 1, so that if you don't specify anything in this field, at least one digit will be printed.

For floating point data of types **f, e**, or **E**, the *precision* field specifies the number of digits to be printed to the right of the decimal point. Trailing 0's will be printed. The default value for *precision* is 6; if you don't specify anything in this field, six digits will be printed to the right of the decimal point. If you specify a *precision* of 0, the decimal point will not be printed unless you specified a # flag.

For floating point data of types **g** or **G**, the *precision* field specifies the maximum number of digits to be printed to the right of the decimal point. If the value is greater than the specified *precision* or less than -4, the value will be printed in exponential format; otherwise, it will be printed in fixed point format. The actual number of digits that are printed depends on the value of the data. The default value for *precision* is 6; if you don't specify anything in this field, six digits will be printed to the right of the decimal point. If you specify a *precision* of 0, the decimal point will not be printed unless you specified a # flag.

For data of type **s** (strings) the *precision* specifies the maximum number of characters to be printed. The default, if you do not specify *precision*, is to print the entire string (until a NULL string-terminating character is encountered).

The *size* Field

The *size* field (if present) changes the size of the data as shown in Table 9-III.

Table 9-II. Using Flags with printf

Flag	Description
None	Right-justify the output if the *width* of the print zone exceeds the number of characters in the value. Pad with zeroes or blanks.
-	Left-justify the output if the *width* of the print zone exceeds the number of characters in the value. Pad with blanks.
+	Print positive values with a leading + character if *type* is **d, i, f, e, E, g**, or **G**.
space	Print positive values with a leading space character if *type* is **d, i, f, e, E, g**, or **G**.
#	Display nonzero octal values (type **o**) with a leading 0. Display nonzero hexadecimal values (type **x** or **X**) with leading 0x or 0X. Display decimal point even if no digits follow it (types **e, E, f, g,** or **G**). Do not remove trailing zeroes (types **g** or **G**). No effect on other types.

Expanded Mandelbrot Set Example

Figure 9-1 lists a program for looking at an expanded section of the Mandelbrot set. The part of the program that draws the picture is the same as the program listed in Figure 8-3, except that ordinary cycling of colors is used instead of the special color selection. What we are concerned with here is the beginning part of the program, which displays a box on the screen that shows the current parameters for the maximum and minimum values used to compute the set. This section of the program uses many different means of displaying characters, numbers, and box characters on the screen. The section begins with a *while* loop that repeats as long as the keyboard entry is not the *Ent* key. (Note that the key value is stored in *ch*, which is initialized to 0, which guarantees that we pass through the loop at least once.) First, we go to the top left corner of our proposed box and use *printf* to print the top

Table 9-III. Size Parameters for printf

Size Character	Used with Types	Effect
h	**d, i, u, o, x, X**	Value has a data type of *short int* or *unsigned short int* rather than *int* or *unsigned int*. Since these are of the same length on the PC, it actually has no effect.
l	**d, i, u, o, x, X**	Value has a data type of *long int* or *unsigned long int* rather than *int* or *unsigned int*.
L	**f, e, E, g, G**	Value has a data type of *long double* rather than *float* or *double*.
F	**p, s**	Pointer value is for a *far* pointer rather than a normal pointer. Used with small or medium memory models.
N	**p, s**	Pointer value is for a *near* pointer rather than a normal pointer. Used with large memory models.

left corner double line box character, *0xC9*. (This shows that it can be done with *printf*, but for a single character *putch* is usually more efficient.) Next, a *for* loop is used to print all of the upper boundary of the box, character by character. Again *printf* is used, making use of the character format *\xCD* for the horizontal double line character. We then use *printf* once more to display the top right double line corner character (using form *\xBB* in the expression). Next, we go down to the bottom left corner of the box and use *putchar* to display the bottom left double line corner character (using the form *\xBC*). Another *for* loop is used to draw the bottom box double line character by character using the *putchar* function and the expression *0xCD*. Then the bottom right corner is drawn with *putchar* and the expression *0xBC*. An additional *for* loop draws the left and right vertical sides of the box using *putch* and the expression *0xBA*. Note that the program is set up so that we are in the text mode at this point, having used *setmode* previously to set the mode to 3. You should try setting the mode to 18 (a graphics mode) at that point, and observe that *putch* no longer works, leaving out the vertical lines of the box. You'll then have to change *putch* to *putchar* to have vertical lines in the graphics mode. Next, we're going to display a heading for the box at the appropriate place on the display. The heading is already stored in the array *legend* and is displayed using the *printf* statement and the type *s* (string). Then four lines are displayed in the box. Each uses *printf* with text string within the function followed by the number of one of the Mandelbrot set bounds in the format *8.6f*, which is a fixed point number having eight digits total and six digits following the decimal point. A final *printf* statement gives instructions on how to change the Mandelbrot set bounds. Next a character is read into *ch* from the keyboard using *getch* which is described in Chapter 8. If the key struck was *1, 2, 3*, or *4*, the boundaries of the Mandelbrot set are changed. The initial bounds are selected by *1*. Each number after that doubles the expansion of the previous set. We then loop through the *while* loop again to permit another selection until the *Ent* key is hit, whereupon the loop terminates and the Mandelbrot set program is run. Plate 7 shows the resulting picture for expansion 4.

Figure 9-1. Listing of Mandelbrot Set Expansion Program Illustrating Use of Character and Number Display to Screen

```
/*
 ╔═══════════════════════════════════════════════════════════╗
 ║                                                           ║
 ║     MANDLBROT = Program to generate Mandelbrot set        ║
 ║                                                           ║
 ║              By Roger T. Stevens   9-10-92                ║
 ║                                                           ║
 ╚═══════════════════════════════════════════════════════════╝
*/

#include <dos.h>
#include <stdio.h>
#include <math.h>

int xres = 640, yres = 480;
int  color, row, col;
double Pmax=-0.745388, Pmin= -0.745408, Qmax = 0.113040,
```

(continued)

```
    Qmin = 0.112976;
double Xmax[4]={-0.745388, -0.745408, -0.745418,
-0.745423};
double Xmin[4]={-0.745408, -0.745448, -0.745438,
-0.745433};
double Ymax[4]={0.113040, 0.113024, 0.113016, 0.113020};
double Ymin[4]={0.112976, 0.112992, 0.113000, 0.113004};
double Q[480], P, deltaP, deltaQ, x, y, xsq, ysq;

void plot(int x,int y,int color);
void setmode(int mode);
void cls(int color);

void main()
{
    int i;
    char ch=0, legend[26]={"MANDELBROT SET GENERATOR"};
    setmode(3);
    while (ch != 0x0D)
    {
        gotoxy(10,4);
        printf("%c",0xC9);
        for (i=0; i<59; i++)
            printf("\xCD");
        printf("\xBB");
        gotoxy(10,14);
        putchar(0xC8);
        for (i=0; i<59; i++)
            putchar(0xCD);
        putchar(0xBC);
        for (i=5; i<14; i++)
        {
            gotoxy(10,i);
            putch(0xBA);
            gotoxy(70,i);
            putch(0xBA);
        }
        gotoxy(26,5);
        printf("%s",legend);
        gotoxy(15,7);
        printf("X minimum: %8.6f",Pmin);
```

(continued)

```
        gotoxy(15,8);
        printf("X maximum: %8.6f",Pmax);
        gotoxy(15,9);
        printf("Y minimum: %8.6f",Qmin);
        gotoxy(15,10);
        printf("Y maximum: %8.6f",Qmax);
        gotoxy(12,12);
        printf("Select expansion 1, 2, 3, or 4, or"
            "'Ent' to run...");
        ch = getch();
        if ((ch==0x31) || (ch==0x32) || (ch==0x33)
            ||(ch==0x34))
        {
            Pmax = Xmax[ch - 0x31];
            Pmin = Xmin[ch - 0x31];
            Qmax = Ymax[ch - 0x31];
            Qmin = Ymin[ch - 0x31];
        }
    }
    cls(7);
    deltaP = (Pmax - Pmin)/xres;
    deltaQ = (Qmax - Qmin)/yres;
    for (row=0; row<yres; row++)
        Q[row] = Qmin + row*deltaQ;
    for (col=0; col<xres; col++)
    {
        P = Pmin + col*deltaP;
        for (row=0; row<yres; row++)
        {
            x = y = xsq = ysq = 0.0;
            for (color=0; ((color<512) && (xsq + ysq
                < 4)); color++)
            {
                xsq = x*x;
                ysq = y*y;
                y = 2*x*y + Q[row];
                x = xsq - ysq + P;
            }
            plot(col, row, (color % 16));
        }
    }
```

(continued)

```
    getch();
}

/*
┌──────────────────────────────────────────┐
│                                          │
│       setmode() = Sets video mode        │
│                                          │
└──────────────────────────────────────────┘
*/

void setmode(int mode)
{
    union REGS reg;

    reg.x.ax = mode;
    int86 (0x10,&reg,&reg);
}

/*
┌──────────────────────────────────────────┐
│                                          │
│      cls() = Clears the screen           │
│                                          │
└──────────────────────────────────────────┘
*/

void cls(int color)
{
     union REGS reg;

     reg.x.ax = 0x0600;
     reg.x.cx = 0;
     reg.x.dx = 0x1E4F;
     reg.h.bh = color;
     int86(0x10,&reg,&reg);
}
```

(continued)

```
/*
      plot() = Plots a point on the screen at a designated
          position using a selected color for 16 color
                            modes.
*/

void plot(int x, int y, int color)
{
    #define graph_out(index,val)  {outp(0x3CE,index);\
                   outp(0x3CF,val);}

    int dummy,mask;
    char far * address;

    address = (char far *) 0xA0000000L + (long)y *
       xres/8L + ((long)x / 8L);
    mask = 0x8000 >> (x % 8);
    outport(0x3CE,mask | 8);
    outport(0x3CE,0x0205);
    dummy = *address;
    *address = color;
    outport(0x3CE,0x0005);
    outport(0x3CE,0xFF08);
}
```

10

Disk Reading and Writing

Your compiler probably already makes provision for storing the source code and compiled version of each C program on disk. This is sufficient if you never create any data or displays in the program that you would like to save. For the more common situation when you want to be able to save and quickly recall data that your program has worked long and hard to produce, you need to know how to write the portions of a C program that open and close a disk file and store data to it. These topics are covered in this chapter.

Opening a Disk File

To work with a disk file in C, you need to have a name that is a pointer to the file. This name is defined in a statement such as

```
FILE *my_file;
```

You'll be using *my_file* to refer to the file throughout your program. Please note, however, that *my_file* is not the file name that will be used by DOS and that will be associated with the file in the disk directory. The *my_file* term is simply a convenient name for use within the C program. Now, we've got a name, but we still have to open an actual file. This is done by the statement

```
my_file = fopen(filename, mode);
```

The statement starts with the name you have assigned for the file *my_file*. This is set equal to the keyword *fopen* with two arguments in parentheses. The first, *filename*, is the actual DOS name of the file, under which it will be stored and listed in the file directory. It can be any combination of characters acceptable to DOS, consisting of no more that eight characters which may optionally be followed by a period and a three-character extension. The argument *filename* may be the pointer to an array where the file name is stored, or you can enclose the actual name in double quotation marks:

```
my_file = fopen("dos_file.ext", mode);
```

The *mode* argument consists of a string of one or more characters enclosed in double quotation marks. Permissible character combinations and their meanings are shown in Table 10-I.

Table 10-I. File Opening Modes

Mode	Description
r	Open file for reading only. File must exist on disk.
w	Create a file for writing. If file exists, it will be overwritten.
a	Open existing file for writing at end of current data or create new file for writing if file doesn't exist.
rb	Open binary file for reading only. File must exist on disk.
wb	Create a binary file for writing. If file exists, it will be overwritten.
ab	Open existing binary file for writing at end of current data or create new file for writing if file doesn't exist.
rt	Open text file for reading only. File must exist on disk.
wt	Create a text file for writing. If file exists, it will be overwritten.

(continued)

Table 10-I. File Opening Modes (cont).

Mode	Description
at	Open existing text file for writing at end of current data or create new file for writing if file doesn't exist.
r+	Open existing file for update (reading and writing).
w+	Create a file for update (reading and writing). If file exists, it will be overwritten.
a+	Open existing file for reading and for writing at end of current data or create new file for update if file doesn't exist.
rb+ or r+b	Open existing binary file for update (reading and writing).
wb+ or w+b	Create a binary file for update (reading and writing). If file exists, it will be overwritten.
ab+ or a+b	Open existing binary file for reading and for writing at end of current data or create new file for update if file doesn't exist.
rt+ or r+t	Open existing text file for update (reading and writing).
wt+ or w+t	Create a text file for update (reading and writing). If file exists, it will be overwritten.
at+ or a+t	Open existing text file for reading and for writing at end of current data or create new file for update if file doesn't exist.

Your compiler may not support all of the modes shown in Table 10-I. Check to make sure before you use one. You may be wondering, if you use the first three modes listed, whether the file you open will be a binary file, a text file, or what. The answer is that this is controlled by the state of a global variable called *_f_mode*. This variable may be set to either of two values; *O_BINARY*, which causes the file to be opened as a binary file; or *O_TEXT*, which causes the file to be opened as a text file. The default mode is the text mode.

Just what is the difference between text and binary files, anyway? These are characteristics of DOS files that will mostly not concern you when writing C programs. The most significant difference is that when a text file encounters the character *Ctrl-z*, it interprets it as the end of file, whereas a binary file keeps chugging along until the end of the file is reached. Thus if you expect to have data in your file that will include a byte that looks like *Ctrl-z* (1AH), you should be sure to use the binary file format. The two file types also differ in the way that they interpret an end of line. If you are using the *putc, fputc, puts, fputs, putw*, or *fwrite* functions to send data to the disk file, the file contents will look exactly the same regardless of whether the mode is text or binary. If you use *fprintf*, there will be minor differences.

Now we can see what a typical statement to open a file looks like:

```
my_file = fopen("dosfile.ext, "rb");
```

This statement will open the file *dosfile.ext* for binary read only and will assign the name *my_file* for use throughout your program. What happens if *dosfile.ext* doesn't exist? Regardless of whether you are attempting to open a file for reading, writing, or appending, if the program cannot find an existing file when one is essential or if it is unsuccessful at creating a new file (maybe your disk or directory is full) a NULL will be returned in *my_file* instead of the pointer to the data stream that is normally returned. The best programming practice is to check this for NULL after the *fopen* process takes place and have some alternate course of action ready in case the file could not be opened (display an error message and quit, for example). If you get the NULL return and then attempt to read or write from the non-existent file, you're in for big trouble. You'll see in the Tchebychev fractal program given below that I've disregarded my own advice. I haven't had any trouble yet, but you may. (I did it to make things as simple as possible in the illustration.)

Closing a File

When you're through with a file, you should close it. This is pretty simple. You simply use the statement

```
fclose (my_file);
```

where *my_file* is the name you assigned to designate the file in the *fopen* statement.

The *fseek* and *rewind* Commands

If you select one of the update modes given above, you will find yourself at the beginning of the file. If you want to read from the file at the beginning, you're in good shape. If you want to write, you need to move to some other point on the file unless you're willing to write right over the current file data. The *fseek* command allows you to position the file pointer at any place you want on the file, so that you can read or write beginning at a desired point. The command takes the form

```
fseek(my_file, offset, whence);
```

The argument *my_file* is the name that you assigned to the file with the *fopen* statement. The argument *offset* is a long integer which indicates the number of bytes that you are to move the file pointer. The argument *whence* determines the point of reference from which you are going to move with the *fseek* statement. If this is set to 0, you'll start at the beginning of the file and move *offset* bytes toward the file end. If *whence* is set to 1, you'll start at the current position of the file pointer and move *offset* bytes. If *whence* is 2, you'll start at the end of the file and move *offset* bytes.

The *rewind* command takes the form

```
rewind(my_file);
```

This returns you to the beginning of the file. Hence, *rewind* is the same as

```
fseek(my_file, 0L, 1);
```

The *putc* and *fputc* Functions

The *putc* function outputs a single character to the data stream. If you use *putc* like this

```
putc(ch,my_file);
```

where *my_file* is the name of the file that you opened using *fopen*, then the character will be sent to the file at the current file pointer location. There is also a function *fputc*, which is exactly the same as *putc* as far as its operation is concerned. The only difference is that *fputc* is a true function, whereas *putc* is a macro that will be inserted into the compiled code at each appropriate location. Normally, in programming, you won't be much concerned about this difference. It will manifest itself by the fact that using *putc* may result in a slightly faster and longer compiled program than using *fputc*.

The *fputs* Functions

For outputting a whole string of characters in one operation, we have the function *fputs*. It is used just like the *fputc* function except that the parameter is the address of a character string, instead of a single character. If you have defined a character string as

```
char string[34];
```

and then loaded this array with a string of 33 characters or less [the array must include a NULL character (00) at the end of the string], then you can output it to your disk file:

```
fputs(ch,my_file);
```

If you want to include the string within the *fputs* statement, you enclose it within double quotation marks rather than the single quotation marks used for the single character function, like this

```
fputs("Welcome to the world of fractals...",my_file);
```

If you are using this technique, you can use *x* to include untypeable characters within the string. This statement

```
fputs("\x01Welcome to fractals...\x01",my_file);
```

will send the sentence with a smiling face (ASCII character 01H) on either side of it to the disk file.

The *putw* Function

The *putw* function is similar to *putc* except that it writes a word (of length integer, or two bytes) to the specified stream. This enables you to output two characters in one operation. If you have defined an integer by

```
int integer;
```

at the beginning of your program or function, you can then send it to a stream (in this case the disk file *my_file*) as follows:

```
putw(integer, my_file);
```

You can also define the two characters to be output within the *putw* statement:

```
putw('HA', my_file);
```

Note that even though there are two characters specified, they are enclosed in single quotation marks as if they were a single character. You can also use *0x* to precede a number representing two ASCII characters:

```
putw(0x4148, my_file);
```

Because of the way in which the PC handles words, the two least significant digits of this four-digit number will appear first in the disk file, followed by the two most significant digits.

The *fprintf* Function

If you want to convert a combination of characters and numerical outputs to ASCII format and then output it to your disk file, you can use the *fprintf* statement. This function works in exactly the same way as the *printf* function, except that the very first argument passed to the function is the name of your file (*my_file*, for example) and the output is sent to the disk file instead of the display screen. Let's look at a couple of examples:

```
fprintf(my_file,"The value of a is %d",a);

fprintf(my_file,"%4.2f   4.2f",dec1,dec2);
```

The first statement sends the ASCII string *The value of a is* to the file followed by the ASCII characters that make up the number stored in *a*. The second statement takes the contents of *dec1* and *dec2* and creates an ASCII string for the value of each that has four digits, with two of them following the decimal point. These strings are then output to the disk file with three spaces separating them.

The *fread* and *fwrite* Functions

If you use any of the commands given previously to generate a disk file in piecemeal fashion, you are likely to have a slow program, since there will be a lot of individual disk accesses, each of which takes a large amount of time. Usually, you can make your program run a lot faster by gathering together all or part of the information that is going to a disk file in a buffer and then using a single *fwrite* call to send it to the disk in one operation. Similarly, you can speed up disk reads by using a single *fread* call to transfer a large chunk of information from the disk to a buffer and then process the data from the buffer for use in your program. The *fwrite* statement works like this:

```
fwrite(buffer,item_size,no_of_items,my_file);
```

The term *buffer* is the name of any array that you have defined in your program that can hold all of the information that you are going to write to disk with the *fwrite* statement. You will have defined it near the beginning of the program with a statement such as

```
char buffer[1600];
```

The term *item_size* indicates the size of the data items stored in *buffer*. If *buffer* is of type *char* as it was just defined, then *item_size* will be 1; if it is of type *int*, the *item_size* will be 2; if it is of type *long*, the *item_size* will be 4 and so forth. The term *no_of_items* is the number of these items of data that are to be written to the disk file by the *fwrite* statement. This can be any number that you want as long as it doesn't exceed the buffer size, so for the example just given, the maximum data transferred would be 1600 items. The actual number of bytes sent to the disk file is *item_size* × *no_of_items*. The final term in the *fwrite* statement is the name that you have assigned to the disk file in the *fopen* statement.

The *fread* statement is just like the *fwrite* statement, having the form

```
fread(buffer,item_size,no_of_items,my_file);
```

The terms all have the same meaning, but in this case, the data is read from the disk file and stored in *buffer*.

The Tchebychev C_5 Polynomial Fractal Curve

The Tchebychev polynomials are another set of orthogonal polynomials found in math reference books which make good fractal curves. The Tchebychev C_5 polynomial has the value

$$C_5 = z^5 - 5z^3 + 5z \qquad \text{(Equation 10-1)}$$

The resulting iterated equation is

$$z_n = c(z_{n-1}^5 - 5z_{n-1}^3 + 5z_{n-1}) \qquad \text{(Equation 10-2)}$$

Figure 10-1 lists a program for generating a fractal using this equation. The resulting picture appears in Plate 8. The program for generating the fractal is very similar to those described in previous chapters for creating similar fractals, so we won't go over that again. What's interesting for this chapter is an added function called *savescrn*, which saves the completed Tchebychev display to a disk file. The file that is created is very simple. It doesn't have any header nor does it make any attempt to compress the data. It therefore is large, 153,200 bytes, which represents four bits for every pixel of the 640 × 480 pixel display screen. The display is read one memory plane at a time. All data for the complete display for the first memory plane is written to the disk file, then that for the second memory plane, and so forth. If you look at the listing for *savescrn*, you'll see that we first define the file *fsave*, and a character buffer of 80 characters, which is what it takes for one line of pixels (640) with 1 bit per pixel. Next, we open *fsave* as a binary file for writing which is to have the name *tcheb.raw* in the DOS directory. The function then begins a *for* loop, which iterates once for each of the four memory planes. Nested within are two more *for* loops. The outer one iterates once for each row of the display. The inner loop iterates once for each column of a row. For each pass through the inner loop, the *getByte* function is called to get a byte of data from the display at the proper location. This byte is then stored in the proper member of the *buffer* array. At the end of each row, an *fwrite* call is made to store the 80 bytes from the disk file to the disk file. Note that instead of storing this data in the buffer, you could take each byte from *getByte* and write it directly to the disk file with an *fputc* function call. You ought to try this as an exercise. You'll find that this way takes much longer to write the disk file than the way given in the listing.

The *getByte* function begins by computing the absolute memory address of the byte that is to be read from the display memory. Remember that this address actually represents four different memory planes that are internal to the VGA. The two *outport* calls set up the VGA registers so that when data is read from the memory plane address, only data from one selected memory plane is actually read. This reading operation takes place when the *return* statement calls for the contents of the selected memory address. At that point the function terminates and the data from the selected memory plane and address is returned to the calling function.

You can add the functions *savescrn* and *getByte* to any of the fractal generating programs given in this book and thereby provide the capability to save the fractal display to disk. In the *fopen* statement, you'll need to assign a unique DOS file name within the double quotes for each time you use the function, to assure that each fractal display is saved in its own file. Alternately, you might want to modify the function to read a file name from the keyboard or to generate a file name in some other way that results in a unique file name for each display.

Figure 10-1. Listing of Program to Generate and Store Tchebychev C_5 Fractals

```
/*

    TCHEBYCHEV = Program to generate Tchebychev C5
                      fractal set.

              By Roger T. Stevens  7-7-92

 */

#include <dos.h>
#include <stdio.h>
#include <math.h>

int xres = 640, yres = 480;
int  color, row, col;
float Pmax = 1.0, Pmin = -1.0, Qmax = 0.4, Qmin = -0.4;
float Q[480], P, deltaP, deltaQ, old_x, old_y, temp_x,
      temp_y, x, y;

void cls(int color);
unsigned char getByte(unsigned long int address, int
      color_plane);
void plot(int x,int y,int color);
void savescrn(void);
void setmode(int mode);

int i;

main()
{
      setmode(18);
      cls(7);
      deltaP = (Pmax - Pmin)/xres;
      deltaQ = (Qmax - Qmin)/yres;
      for (row=0; row<yres; row++)
            Q[row] = Qmax - row*deltaQ;
      for (col=0; col<xres; col++)
```

(continued)

```
        {
                P = Pmin + col*deltaP;
                for (row=0; row<yres; row++)
                {
                        x = P;
                        y = Q[row];
                        old_x = old_y = 0.0;
                        for (color=0; ((color<64) && ((x*x +
                                y*y) < 1000)); color++)
                        {
                                temp_x = x*x*x*x*x -
                                        10.0*x*x*x*y*y +
                                        5.0*x*y*y*y*y - 5.0*x*x*x +
                                        15.0*x*y*y + 5.0*x;
                                temp_y = 5.0*x*x*x*x*y -
                                        10.0*x*x*y*y*y +
                                        y*y*y*y*y - 15.0*x*x*y +
                                        5.0*y*y*y + 5.0*y;
                                x = P*temp_x - Q[row]*temp_y;
                                y = Q[row]*temp_x + P*temp_y;
                                if ((x == old_x) && (y == old_y))
                                {
                                        color = 0;
                                        break;
                                }
                                if ((color % 8) == 0)
                                {
                                        old_x = x;
                                        old_y = y;
                                }
                        }
                        plot(col, row, (color % 16));
                }
        }
        savescrn();
        getch();
}
```

(continued)

```
/*
        setmode() = Sets video mode
*/

void setmode(int mode)
{
      union REGS reg;

      reg.x.ax = mode;
      int86 (0x10,&reg,&reg);
}

/*
        cls() = Clears the screen
*/

void cls(int color)
{
      union REGS reg;

      reg.x.ax = 0x0600;
      reg.x.cx = 0;
      reg.x.dx = 0x1E4F;
      reg.h.bh = color;
      int86(0x10,&reg,&reg);
}

/*
  plot() = Plots a point on the screen at a designated
      position using a selected color for 16 color
                   modes.
```

(continued)

```
*/

void plot(int x, int y, int color)
{
      #define graph_out(index,val)  {outp(0x3CE,index);\
                              outp(0x3CF,val);}

      int dummy,mask;
      char far * address;

      address = (char far *) 0xA0000000L + (long)y *
            xres/8L + ((long)x / 8L);
      mask = 0x80 >> (x % 8);
      graph_out(8,mask);
      graph_out(5,2);
      dummy = *address;
      *address = color;
      graph_out(5,0);
      graph_out(8,0xFF);
}

/*
╔══════════════════════════════════════════════════════╗
║                                                      ║
║        getByte() = Reads a byte from the screen      ║
║                                                      ║
╚══════════════════════════════════════════════════════╝
*/

unsigned char getByte(unsigned long int address, int
color_plane)
{
      char far *video_address;

      video_address = (char far *) 0xA0000000L + address;
      outport(0x3CE,(color_plane<<8) + 4);
      outport(0x3CE,5);
      return (*video_address);
}
```

(continued)

```
/*
 ┌──────────────────────────────────────────────────────┐
 │                                                      │
 │     savescrn() = Saves a graphics screen to disk     │
 │                                                      │
 └──────────────────────────────────────────────────────┘
*/

void savescrn(void)
{
      FILE *fsave;
      char buffer[80];
      int col, i, row;

      fsave = fopen("tcheb.raw","wb");
      for (i=0; i<4; i++)
      {
            for (row=0; row<yres; row++)
            {
                  for (col=0; col<xres/8; col++)
                  {
                        buffer[col] = getByte(80*row +
                              col, i);
                  }
                  fwrite(buffer,1,80,fsave);
            }
      }
      fclose(fsave);
}
```

Restoring a Display from a Disk File

Once you have saved a display with *savescrn* you need a way to recover the display from the disk file. The function *readscr* will perform this task. If you've used other programs to recover displays from disk files, you're going to be amazed at the speed with which *readscr* generates a display. It makes all the other programs seem like snails by comparison. There are three reasons for this speed. First, the program reads the data for a complete memory plane into the buffer with a single *fread* call. Second, the data for a complete memory plane is contiguous in the file and is not compressed, so it can be read directly and quickly. Third, the complete data for each memory plane is transferred from the buffer to display memory with a single *movedata* instruction. You'll learn more about moving data in memory in Chapter 16. The *readscr* program is listed in Figure 10-2. Once the buffer and variables are defined, with the buffer large enough (38,400 bytes) to hold an entire memory plane, the display is set to mode 18 (640 × 480 pixels × 16 colors) and the segment and offset values for the beginning of the buffer are obtained. The file is opened for reading and then a *for* loop begins, which iterates for each of the four color planes. At each iteration, an *fread* call loads the entire buffer with data from the file. Then a register on the VGA card is set to limit communication to a single memory plane. The data is then transferred in one operation using the *movedata* statement to the display memory plane selected. The four planes are loaded so fast that you can barely see the partial displays that are created. You can use this program to redisplay any fractal that you have saved with the *savescrn* function. You just need to make sure that you have the right file name between the double quotes in the call to *fopen*.

Figure 10-2. Program to Restore a Display from a Disk File

```
/*
╔══════════════════════════════════════════════════════════╗
║                                                          ║
║     READSCR = Program to read a graphics information     ║
║          disk file and display it on the screen.         ║
║                                                          ║
║               By Roger T. Stevens  7-7-92                ║
║                                                          ║
╚══════════════════════════════════════════════════════════╝
*/

#include <dos.h>
#include <stdio.h>
#include <math.h>

int xres = 640, yres = 480;
int  color, row, col;

void plot(int x,int y,int color);
void setmode(int mode);
unsigned char getByte(unsigned long int address, int
```

(continued)

```
color_plane);

struct SREGS segregs;
FILE *fsave;
char buffer[38400];
int i, srcseg, srcoff;

main()
{
      setmode(18);
      segread(&segregs);
      srcseg = segregs.ds;
      srcoff = (int) buffer;
      fsave = fopen("tcheb.raw","rb");
      for (i=0; i<4; i++)
      {
            fread(buffer,1,38400,fsave);
            outport(0x3C4,(0x01<<(8+i)) + 2);
            movedata(srcseg,srcoff,0xA000,0,38400);
      }
      fclose(fsave);
      getch();
}

/*
+----------------------------------------------------------+
|                                                          |
|              setmode() = Sets video mode                 |
|                                                          |
+----------------------------------------------------------+
*/

void setmode(int mode)
{
      union REGS reg;

      reg.x.ax = mode;
      int86 (0x10,&reg,&reg);
}
```

11

Keyboard Input

One of the most interesting problems that you'll encounter in writing C programs is how to make sure that you get the correct keyboard input from the user. If you assume that the operator is going to type in exactly what the program asks for, you're going to be very disappointed. Somebody is going to come up with a combination of keystrokes you haven't even imagined that is going to cause your whole program to go completely bananas. C has several primitive functions for reading keystrokes as well as one more sophisticated one, but these aren't, by themselves, capable of protecting against error. In this chapter, we'll begin by discussing these functions. Once you are well grounded in how they work, we'll talk about errors that can occur and how you can protect your program against them.

The *getch* and *getche* Functions

The first function that we shall consider is *getch*. This function reads a character from the keyboard but does not echo it to the screen. The function returns data of type *int*. Since the data returned from a keyboard is always of character length, you can store it in *char* type data. Thus either of the following will work:

```
char ch;
ch = getch();
```

or

```
int ch;
ch = getch();
```

Some of the special characters return two bytes to the keyboard buffer. The first of these is a NULL (00) to indicate that a special character is involved and the second is a character designation, which may be the same as one of the alphanumeric keys. You handle the situation like this

```
ch = getch();
if (ch == 0x00)
    ch = getch();
```

What you do next is up to you. You may have a whole section of code that is included in the bounds of the *if* statement and that is based on the knowledge that special characters only are involved. Another possibility is to add 256 to the special character codes so that they will all be different from ordinary alphanumerics.

The *getche* function is exactly the same as the *getch* function as far as retrieving characters from the keyboard is concerned. The difference is that the *getche* function echoes any key struck to the display screen, with the character appearing at the current cursor location. Unfortunately, *getche* works only with text display modes, so that if you are in a graphics mode, as is the case with most of the programs that we use in this book, *getche* will not display a character so you will have to find another means of showing what you enter at the keyboard on the screen.

The *getc* and *fgetc* Functions

The *getc* function is the same as the *getch* function except that it has a second parameter, which is the stream from which the character is read. If you define the stream as *stdin*, which is normally the keyboard, you'll get the same result as if you were using *getch*. The code would be

```
ch = getc(stdin);
```

You can also use this function to read from a disk file.

The function *fgetc* is exactly the same as *getc* as far as its operation is concerned. The only difference is that *fgetc* is a true function, whereas *getc* is a macro that will be inserted into the compiled code at each appropriate location. Normally, in program-ming, you won't be much concerned about this difference. It will manifest itself by the fact that using *getc* may result in a slightly faster and longer compiled program than using *fgetc*.

The *getchar* and *fgetchar* Functions

The *getchar* function has no parameters. It works like this

```
ch = getchar();
```

It is the same as *getc* with the stream defined as *stdin*. There is also a corresponding *fgetchar* function.

The *gets* and *fgets* Functions

If you want to read a whole string of characters in one operation, you can use the *gets* or *fgets* function. The *gets* function works like this

```
char string[80];
gets(string);
```

where *string* is the address of a buffer where your string is going to be stored. Characters are read from *stdin* (the keyboard until a new line character is encountered (when you hit the *Enter* key). The new line is replaced by a NULL (00) in the buffer.

The *fgets* function has three parameters. The first is the string buffer address as for *gets*. The second is an integer that indicates the number of characters in the string. The third parameter is the name of the stream from which characters are read. If you make this *stdin* the function will read from the keyboard. You can also use it to read disk files. The function reads characters to the buffer until it has read one less than the number of characters specified or until a new line character is encountered. It then appends a NULL (00) as the last character of the string in the buffer. The code is

```
char string[80];
fgets(string,80,stdin);
```

The *getw* Function

The *getw* function is similar to *getc* except that it reads a word (of length integer, or two bytes) from the specified stream. If used to read from a disk file, the file must be opened in the byte mode. The code is

```
int integer;
integer = getw(stdin);
```

The *scanf* Function

The *scanf* function will read a series of input fields from *stdin*, one character at a time. The *scanf* function provides a lot of flexibility in formatting and defining numbers and strings that are to be read in from the keyboard. (It automatically inputs from *stdin*.) The *scanf* function can handle a variable number of parameters. The first parameter is always a format string, which is enclosed in double quotation marks. The remaining parameters are **address of arguments** whose values will be read in the form specified in the format string. This is unlike *printf*, where the argument values, rather than their addresses, are passed to the function. You, as programmer, need to make sure that your format string specifies the format of each variable in your parameter list and that you have a variable in the list for each format specifier. If the number of format specifiers and the number of variables do not match, you can get some really strange results from the *scanf* function.

Unlike the *printf* function, where anything in the format string that is not a format specifier is printed out, the *scanf* function format string ignores all format specifiers, so it should contain nothing else.

Specifying the Input Format

The entire format specifier used to determine the format of an input is of the form

% [*] [width] [F | N] [h | l | L] type

Table 11-I indicates the meaning of these symbols. Note that the ones in brackets are optional and may be used or not as the occasion demands.

The straight lines separating characters indicate that any one of the characters may be used at that point. For each variable in the variable list of your *scanf* statement, you must have a format specifier in the format string. As a minimum, the format specifier must consist of a % character followed by a *type* character. It may also include any of the optional characters just described.

Table 11-I. Format Specifier Symbols for scanf

Format Specifier	**Meaning**
*	Suppresses the next input.
width	Maximum number of characters to be read. Since *scanf* terminates a field when it encounters a space, a carriage return, or an unconvertible character, the actual number of characters read may be less.
N	Overrides default size of address argument and makes it a near pointer.
F	Overrides default size of address argument and makes it a far pointer.
h	Overrides the default type of address argument and makes it a *short int*.
l	Overrides the default type of address argument. If an integer was specified, it is converted to a *long int*. If a floating point number was specified, it is converted to a *double*.
L	Overrides a default type of floating point and makes it a *long double*.

Type Characters

The *type* character determines the format in which the output is displayed. Table 11-II lists all of the *type* characters and what is read to the argument when each is invoked. Make sure you use a *type* character that corresponds to the type of argument that you are to be processing; otherwise the results will be strange and unpredictable. Note that each *type* has a default precision that will be used to format the input data if you do not specify otherwise.

Table 11-II. Type Characters Used with scanf Statement

Type	Input Type	Type of Argument
c	Character.	Pointer to *char* (or array of *char* if width is greater than one).
d	Decimal integer.	Pointer to *int*.
D	Decimal integer.	Pointer to *long*.
e,E	Floating point.	Pointer to *float*.
f	Floating point.	Pointer to *float*.
g,G	Floating point.	Pointer to *float*.
i	Decimal, octal, or hexadecimal integer.	Pointer to *int*.
l	Decimal, hexadecimal, or octal integer.	Pointer to *long*.
n		Pointer to *int*.
o	Octal integer.	Pointer to *int*.
O	Octal integer.	Pointer to *long*.

(continued)

Table 11-II. Type Characters Used with printf Statement (cont.)

Type	Input Type	Type of Argument
p	Hexadecimal form *xxxx:yyyy*, where *xxxx* is segment address and *yyyy* is offset.	Pointer to an object (*far** or *near**).
s	Character string.	Pointer to array of *char*.
u	Unsigned decimal integer.	Pointer to *unsigned int*.
U	Unsigned decimal integer.	Pointer to *unsigned long*.
x	Hexadecimal integer.	Pointer to *int*.
X	Hexadecimal integer.	Pointer to *long*.
%	% character.	% is stored.

An Example Using *scanf*

In Chapter 10, we showed how to save a display to disk and then to read the disk file back to a display. The disk file was named *tcheb.raw* and this name was built into the *readscr* program. Here, we're going to generalize the program for reading a disk file to screen using *scanf* to allow you to type in the name of any disk file that you want. To make the program work, you need to have already generated the display data file *tcheb.raw* or some other display data file in the same format, and then type in the appropriate name when queried by the program. The program, which is called *readscr2*, is listed in Figure 11-1. Note how simple it is to read in the file name. First we use *printf* to give directions to the user. Then we use *scanf* to read in the keystrokes to *filename*, which is an array of 14 characters. This works just fine if you type in the proper file name.

Figure 11-1. Program to Read and Display a User-Selected File

```
/*
┌──────────────────────────────────────────────────────────────┐
│                                                              │
│   READSCR2 = Program to read a graphics information          │
│        disk file and display it on the screen.               │
│                                                              │
│            By Roger T. Stevens  7-16-92                      │
│                                                              │
└──────────────────────────────────────────────────────────────┘
*/

#include <dos.h>
#include <stdio.h>
#include <math.h>

int xres = 640, yres = 480;
int  color, row, col;

void setmode(int mode);
void cls(int color);
unsigned char getByte(unsigned long int address, int
color_plane);

struct SREGS segregs;
FILE *fsave;
char buffer[38400], filename[14];
int i, srcseg, srcoff;

main()
{
      setmode(18);
      segread(&segregs);
      srcseg = segregs.ds;
      srcoff = (int) buffer;
      gotoxy(20,6);
      printf("Enter name of file to display: ");
      scanf("%s",filename);
      fsave = fopen(filename,"rb");
      for (i=0; i<4; i++)
      {
```

(continued)

```
                fread(buffer,1,38400,fsave);
                outport(0x3C4,(0x01<<(8+i)) + 2);
                movedata(srcseg,srcoff,0xA000,0,38400);
        }
        fclose(fsave);
        getch();
}

/*
┌──────────────────────────────────────────────────────────┐
│                                                          │
│               setmode() = Sets video mode                │
│                                                          │
└──────────────────────────────────────────────────────────┘
*/

void setmode(int mode)
{
        union REGS reg;

        reg.x.ax = mode;
        int86 (0x10,&reg,&reg);
}
```

Problems with *scanf*

If you're a fairly experienced computer user, you could use the program just described for a long time and never have any problems. However, it doesn't provide any protection against dumb mistakes. If you're trying to write commercial software, you'll need to make provision for every idiotic thing that an inexperienced user might do. There are three basic things that might be done wrong with the *readscr2* program. First, the user might hit the *Enter* key before he starts to type in the file name. In this case, *scanf* will still keep waiting for the specified string to be input (it won't buy off on a null length string) but it will use the *Enter* to move the cursor from its current position to the beginning of the next line. (If you're trying to write within a box, the cursor may suddenly be outside the box altogether.) If you're in a graphics mode, the cursor isn't shown on the screen, so you may not be sure exactly where you are. The second problem is that the user may type in a filename that does not exist. The third problem is that the user may type in a filename that contains characters that are illegal for a DOS filename or type a filename that is longer than allowed. What happens in either of these cases may differ, depending on your compiler. With Borland C++, when an attempt is made to open the specified file, the pointer to the file location will be returned as a NULL. When you attempt to read this file, the screen will go blank. If you then strike any key, the error message

Null pointer assignment

will be displayed and the program will terminate. As programmer, you have to decide how you are going to prevent this sort of problem. One of the most foolproof methods, which won't be shown in detail here, is to write your software to show a directory of all files that are acceptable for use with this program and let the user select the one he wants with the cursor arrows. This assures that only correct program names will be given, but it is the most complicated to implement. Figure 11-2 shows another approach. First, observe that we do not use *scanf*. The method used is more complicated, but prevents the possibility of getting the cursor in the wrong place. It begins with a *for* loop that specifies no parameters. We haven't encountered this before; what it does is to loop forever, with no initialization, no test for ending, and no changing of any argument. The only way to leave such a loop is by encountering a *break* statement. Within this loop we first set the argument *i* to 0 and then start a *while* loop that reads characters into the *filename* array, incrementing the index at each iteration, and continues this until the *Enter* key (0x0D) is hit. Within the loop, each character is displayed on the screen. If 12 characters have been read without encountering the *Enter* key, we break out of the loop, since this is the maximum allowable size for a DOS file name. When we leave the file loop, either the 0x0D character was encountered before the maximum number of character inputs occurred, in which case we replace it with a NULL to terminate the string properly, or the maximum number of characters were input, in which case we place the NULL after the 12 characters. We next attempt to open the specified file. If a matching file cannot be found, then *fsave* contains a NULL. If this is not the case, we have successfully opened the file, so we break from the infinite *for* loop and continue with the rest of the program. If the file wasn't opened successfully, the program goes back to the display line that asked the user to enter a file name and blanks it out through a *for* loop that writes out 80 spaces to the screen. (We use *putchar* since we are in the graphics mode.) Then this same display line is reprinted with

Unable to locate filename nnnnnnnn...Please try again

The *nnnnnnnn* is the file name as the user actually typed it in. The program then does another iteration of the loop to read in another file name. This continues until an acceptable file name is entered. Note that this doesn't address the problem where the user finds that he has no idea of an acceptable file name to type in. In that case, he can't get out of the loop to terminate the program. This approach also doesn't give the user any clues as to why his file name is incorrect. Hopefully, he will look at what is displayed on the screen and identify the error so that he can type in the correct name on the next iteration.

Figure 11-2. Program to Read and Display a User Selected File with Protection Against Errors

```
/*

     READSCR4 = Program to read a graphics information
          disk file and display it on the screen
               with error protection.

               By Roger T. Stevens  7-25-92

*/

#include <dos.h>
#include <stdio.h>
#include <math.h>

int xres = 640, yres = 480;
int  color, row, col;

void setmode(int mode);

struct SREGS segregs;
FILE *fsave;
```

(continued)

```
char buffer[38400], filename[14];
int i, srcseg, srcoff;

main()
{
      setmode(18);
      segread(&segregs);
      srcseg = segregs.ds;
      srcoff = (int) buffer;
      gotoxy(5,6);
      printf("Enter name of file to display: ");
      for (;;)
      {
            i=0;
            while((filename[i++] = getch()) != 0x0D)
            {
                  putchar(filename[i-1]);
                  if (i == 12)
                        break;
            }
            if (filename[i-1] == 0x0D)
                  filename[i-1] = NULL;
            else
                  filename[i] = NULL;
            fsave = fopen(filename,"rb");
            if (fsave != NULL)
                  break;
            gotoxy(1,6);
            for (i=0; i<80; i++)
                  putchar(' ');
            gotoxy(5,6);
            printf("Unable to locate filename"
                  "'%s'...Please try again ",
            filename);
      }
 for (i=0; i<4; i++)
      {
            fread(buffer,1,38400,fsave);
            outport(0x3C4,(0x01<<(8+i)) + 2);
            movedata(srcseg,srcoff,0xA000,0,38400);
      }
```

(continued)

```
        fclose(fsave);
        getch();
}

/*
┌──────────────────────────────────────────────────────────────┐
│                                                              │
│                  setmode() = Sets video mode                 │
│                                                              │
└──────────────────────────────────────────────────────────────┘
*/

void setmode(int mode)
{
        union REGS reg;

        reg.x.ax = mode;
        int86 (0x10,&reg,&reg);
}
```

The *bioskey* Function

The functions that we have encountered thus far in this chapter read characters from the keyboard that depend upon which keys are struck. Usually, the character or characters returned by a keystroke are all the information that we need to know. Sometimes, however, this is not the case. Suppose, for example, that we want to use the arrow keys on the numeric keypad to move a cursor around on the screen in fine gradations and that we also would like to use the Shift-arrow keys to move the cursor in coarse gradations. If we use any of the functions just described, we'll get back ordinary numbers from the shifted arrows, which we can't distinguish from the other number keys. Fortunately, there is a function called *bioskey* that returns all of the information produced by a keystroke. The data is returned in the form of an integer where the least significant byte of the integer is the ASCII character produced by the key and the most significant byte is the key identification code. Together, these produce a totally unambiguous description of the key information. Figure 11-3 lists a test program for checking out the *bioskey* function. The function begins with a *while* loop, which uses the *bioskey* function to read the key information into the integer *c*. The loop continues as long as the key struck is not the *Esc* key; when this key is struck, the loop terminates. Within the loop, the character code is first broken off from *c* and stored in *character*. If *character* is a printable character, it is also stored in *char2*; otherwise a space is stored in *char2*. (This is done so that we don't try to display a nonprintable character and as a result do something to the display that we didn't want to do.) The key scan code is broken off from *c* and stored in *scancode*. Next, the screen is cleared and a print out of the key information is produced that consists of the hexadecimal representation of the ASCII character, the character itself, if it is printable, the key scan code, and the numerical representation of *c* itself. This continues for every key that is struck until the *Esc* key is encountered. If you are going to use some of this key information in a program, you can use this test program to obtain the information you need for the keys in which you are interested.

Figure 11-3. Test Program for bioskey Function

```
/*

     BIOSKTST = Program to read and display keyboard
                    inputs with bioskey

              By Roger T. Stevens   7-17-92

*/

#include <stdio.h>
#include <dos.h>
#define ESCKEY 0x011B

void setmode(int mode);

void main(void)
{
      char scancode, character, char2;
      int c=0;

      setmode(18);
      while ((c = bioskey(0)) != ESCKEY)
      {
```

(continued)

```
            character = c & 0x00FF;
            if (isprint(character))
                  char2 = character;
            else
                  char2 = ' ';
            scancode = c >> 8;
            gotoxy(1,1);
            clrscr();
            printf("character: %3x  (%2c)  key scan code: %3x"
                  " return data: %d",
                  character,char2,scancode,c);
      }
}

/*
╔══════════════════════════════════════════════════════════╗
║                                                          ║
║               setmode() = Sets video mode                ║
║                                                          ║
╚══════════════════════════════════════════════════════════╝
*/

void setmode(int mode)
{
      union REGS reg;

      reg.x.ax = mode;
      int86 (0x10,&reg,&reg);
}
```

The Mandelbrot Set as a Map of Julia Sets

The Mandelbrot set is known to serve as a map of Julia sets. Moving around on the periphery of the Mandelbrot set, interesting Julia sets are found where the coordinates of *c* in the Julia equation correspond to points on the Mandelbrot set where a cusp or inflection point occurs. The program to be described here is the most complicated we have encountered so far, but it is mainly made up of much simpler programs that have been described earlier. It is designed as an illustration of the use of the *bioskey* function, but also has a number of other interesting features that will teach you some interesting aspects of C programming. The program divides the screen into quadrants and first paints the Mandelbrot set in the top left quadrant. A cursor arrow then appears on the Mandelbrot set. The program then begins drawing a quick outline in the second quadrant of the Julia set associated with the cursor location point on the Mandelbrot set and displays instructions and the coordinate values in the third quadrant. You can move the cursor about until you like the looks of the Julia set outline. You then hit the *Enter* key to cause the program to paint the complete Julia set in the fourth quadrant. When the Julia set is done, you can either hit the *Enter* key to leave the program or hit any other key to allow you to reposition the cursor and draw another Julia set. A typical resulting display is pictured in Plate 9. The program is listed in Figure 11-4.

The program begins with a number of definitions. The first two of these are the scale factor and the maximum size. These are used in the set operations to permit integer arithmetic. This will be explained later. The rest of the definitions define the numbers that are returned when using the *bioskey* function to represent various keys. These were obtained from the test program just described. The main program begins by setting the graphics mode and clearing the screen. It then uses several *for* loops to draw a three-pixel-wide line to box in the screen and then two lines dividing it into four quadrants. The program then runs the function *mandel* to draw the Mandelbrot set in the top left quadrant. Next, the program calls the function *move_cursor*. As long as this function does not read an *Ent* key from the keyboard, it continues to read keystrokes and move the cursor arrow accordingly around the Mandelbrot set. Between keystrokes, it draws a Julia set outline in the top right quadrant. When the *Ent* key is finally encountered, the function returns to the main program, putting the coordinates of the latest cursor arrow location in *julia_p*. The main program then calls the function *julia* with these coordinates to draw the Julia set in the bottom right quadrant. When this is complete, additional instructions are displayed in the bottom left quadrant. When a key is hit, these instructions are blanked and if the key struck was not the *Ent* key, another iteration of the *while* loop occurs. Otherwise, the program terminates.

The Mandelbrot Set with Integer Arithmetic

If you don't have a math coprocessor, you've probably observed that generating the Mandelbrot set or any of the other fractal curves in this book is a lengthy process. We want this program to run fast enough to be useful, so we've written the Mandelbrot set function to work with integer arithmetic rather than floating point. Even using long integers, this is much faster than floating point operations. (By going to assembly language, you could obtain a lot more speed, but that's not something we want to get into in a beginner's book.) The function begins by multiplying the usual limits for the Mandelbrot fractal by a scale factor of 4,194,304. It then computes the increments for each pixel on the screen in these units. It then enters three nested *for* loops. The first two loops cover every pixel in the top left quadrant of the screen, except for the line boundaries of the quadrant that were drawn earlier. The innermost loop iterates the repeated Mandelbrot equation until 64 iterations have taken place or until the sum of the squares of the real and imaginary parts exceeds *MAXSIZE*, which is four times the scale factor. We're next going to use some mathematical equations that are very much like the ones used in the original Mandelbrot set program given in Chapter 7. The difference is that we are using integers, with an imaginary decimal point that we must keep track of to make the results come out right. If we're adding or subtracting two of our numbers, the imaginary decimal point is unchanged, so we have no problem. If we multiply two numbers, we end up with twice as many decimal places as we had before and need to lop off some decimal places to get the new number back in alignment with the old ones. To do this, we establish *xt* and *yt,* which are *x* and *y*, respectively, each shifted 11 decimal bits to the right. When these numbers are multiplied with each other, the result has the imaginary decimal point properly aligned with our original numbers. Since the Mandelbrot set is symmetrical around the *x* axis, one other thing we have done to speed things up is to plot two pixels, one at the top and one at the bottom, for each set of computations. This reduces our computation time to one-half.

Moving the Cursor and Drawing the Preview Julia Set

The function *move_cursor* is the most complicated one in the program. It begins by initializing the argument *zero_buf*, which will be used to blank lines of the top right quadrant of the display, while preserving the white border at the ends. Next, it sets up the value of *c* in the Julia equation to correspond to the cursor position. It then displays information on how to continue and the value of *c* in the bottom left quadrant. Now, if the argument *flag* is 0, the function *arrow* is called to display an arrow at the cursor position. (If *flag* is 1, the function has already run at least once, so there already is an arrow displayed on the Mandelbrot set and its coordinates are in *c_col* and *c_row* since these are static variables, meaning that they are not reinitialized each time the function is run.) The function next enters a *while* loop that continues to iterate until the *Ent* key is pressed. Several arguments are initialized. The function then enters another *while* loop. This loop continues as long as the function *kbhit* is not true. This function keeps checking to see whether a key has been struck. If it does detect that a key was struck, it returns a *True* but doesn't otherwise affect the key information, which remains available to be read by some other function. The remaining code within this *while* loop generates points to create an outline of the Julia set whose value of *c* corresponds to the cursor arrow position. This is done by solving the iterated equation backward instead of forward. Regardless of the starting point, after a few iterations, all points that are calculated fall on the boundary of the Julia curve. You'll note that we then plot each point if it is within the bounds of the top right quadrant and if it is not one of the first 20 points.

Once a keystroke is detected, we leave this inner *while* loop and read the key with the *bioskey* function. The *arrow* function is then run again. Since the arrow was plotted originally by exclusive-ORing the data with the original display, another exclusive-OR at the same point restores the original display and the arrow vanishes. The function then enters a *switch* statement that checks all of the arrow and shift-arrow key codes. For any regular arrow key, the cursor arrow coordinates are moved one pixel in the arrow direction; for any shift-arrow key, the cursor arrow coordinates are moved five pixels in the arrow direction. This movement occurs only if it can without the arrow point going outside the quadrant boundaries; otherwise the keystroke is ignored. After the *switch* statement has caused the cursor arrow coordinates to be changed, the *arrow* function is called again to display the arrow at its new position. The coordinates of *c* of the Julia equation are then updated to match the new cursor position. Next, if the argument *ch* is not that of the *Ent* key, the whole top right quadrant is cleared to be ready for a new Julia outline. (If the *Ent* key was struck, the display is left in the quadrant while the program goes on. Finally, if the *Ent* key was struck, the loop ends and the function ends, returning the coordinates of *c*.

The *arrow* Function

The *arrow* function simply makes use of two *for* loops and a number of *plotxor* statements to exclusive-OR the pattern of an arrow to the display screen. The point of the arrow is the x and y coordinates that are passed to the function.

The *julia* Function

The *julia* function creates the final version of the Julia set in the bottom right quadrant. It uses the same method of integer arithmetic that we described previously for the Mandelbrot set. You'll observe three main differences. First, the column and row starting and ending points are different so that we draw in the bottom right quadrant instead of the top left quadrant. Second, instead of having a new value for P and Q for each pixel, we use the constants p and q that do not change throughout the drawing of the picture and which are passed as arguments to the function. Third, we use a technique for setting the color that chooses a color based on the value that c settles to when the equation does not blow up on iteration. All pixels that do blow up are assigned the background color of dark blue.

The *gotoXY* Function

Borland C++ has a function called *gotoxy* which we have used before. It is actually designed for standard text modes, but we have used it with graphics modes. If you are using a different C compiler and found that it lacked that function, here is a substitute function *gotoXY* which is the answer to your problems. We introduce it here because the Borland function won't go beyond row 25, which is the bottom row for standard text. Mode 18, the 640 × 480 pixel × 16 color graphics mode that we are using has 30 lines of text and we want to use one below row 25. Since the Borland *gotoxy* function won't hack it, we've designed our own using the ROM BIOS video services directly. It's pretty simple, and we've doctored it to give the same results as *gotoxy* when it calls for a position within the first 25 lines, but this function will go up to line 30 with no trouble.

The *plot* Function

The plot function that we've used here looks a little different because we've used the *outport* function directly instead of defining our own output expressions. If you compare it closely with the versions of the *plot* function previously given, you'll see that it is exactly the same.

The *plotxor* Function

This function, instead of overwriting a pixel on the screen with new color information, exclusive-ORs the old pixel data with the color white. This requires use of the sequence registers, so the function is a little more complicated than the ordinary *plot* function. The advantage of this function is that a second application of the function to the same location returns that location to the original color. Thus a cursor or other pattern can be erased by simply rewriting it at the same location.

Figure 11-4. Listing of Mandelbrot-Julia Generation Program

```
/*

    MANDJULIA = Program to pick Julia sets from the
                    Mandelbrot set

           By Roger T. Stevens  7-24-92

*/

#include <dos.h>
#include <stdio.h>
#include <math.h>
#include <stdlib.h>
#include <conio.h>

#define SCALE                         4194304L
#define MAXSIZE                      16777216L
#define LEFT_ARROW                      19200
```

(continued)

```
#define RIGHT_ARROW                    19712
#define UP_ARROW                 18432
#define DOWN_ARROW                     20480
#define SHIFT_LEFT_ARROW               19252
#define SHIFT_RIGHT_ARROW              19766
#define SHIFT_UP_ARROW           18488
#define SHIFT_DOWN_ARROW               20530
#define ENTER                           7181

union REGS reg;

int xres = 640, yres = 480;
int  color, row, col;
typedef struct
{
      float cx;
      float cy;
} Julia;
Julia julia_p;

void arrow (int x, int y);
void gotoXY(int col, int row);
Julia move_cursor(void);
void plot(int x,int y,int color);
void plotxor(int x, int y);
void setmode(int mode);
void cls(int color);
void mandel(void);
void julia(long int p, long int q);

void main()
{
      int i,j;
      int ch;

      setmode(18);
      cls(0);
      for (i=0; i<2; i++)
            for (j=0; j<640; j++)
            {
                  plot(j,i,15);
```

(continued)

```
                plot(j,479-i,15);
                plot(j,240-i,15);
            }
        for (i=0; i<2; i++)
            for (j=0; j<480; j++)
            {
                plot(i,j,15);
                plot(639-i,j,15);
                plot(320-i,j,15);
            }
        mandel();
        while (ch != ENTER)
        {
            julia_p = move_cursor();
            julia(julia_p.cx*4194304L,
                julia_p.cy*4194304L);
            gotoXY(3,26);
            printf("Hit 'Enter' to quit;");
            gotoXY(3,27);
            printf("Any other key for a new set...");
            ch = bioskey(0);
            gotoXY(3,26);
            printf("                    ");
            gotoXY(3,27);
            printf("                              ");

        }
}

/*
╔══════════════════════════════════════════════════════════╗
║                                                          ║
║         mandel() = Generates the Mandelbrot set          ║
║                                                          ║
╚══════════════════════════════════════════════════════════╝
*/

void mandel()
{
        int i;
        long int Pmax, Pmin, Qmax, Qmin;
        long int Q[126], P, deltaP, deltaQ, x, y, xt, yt,
```

(continued)

```
        oldx, oldy;

    Pmax = 1.2 * SCALE;
    Pmin = -2.2 * SCALE;
    Qmax = 1.3 * SCALE;
    Qmin = -1.3 * SCALE;
    deltaP = (Pmax - Pmin)/(xres/2);
    deltaQ = (Qmax - Qmin)/(yres/2);
    for (row=0; row<126; row++)
        Q[row] = Qmin + row*deltaQ;
    for (col=3; col<319; col++)
    {
        P = Pmin + col*deltaP;
        for (row=3; row<122; row++)
        {
            x = P;
            y = Q[row];
            xt = yt = oldx = oldy = 0;
            for (color=0; ((color<64) && ((xt*xt +
                yt*yt)<MAXSIZE)); color++)
            {
                xt = (x>>11);
                yt = (y>>11);
                x = xt*xt - yt*yt + P;
                y = 2*xt*yt + Q[row];
                if ((x == oldx) && (y == oldy))
                {
                    color = 0;
                    break;
                }
                if ((color % 8) == 0)
                {
                    oldx = x;
                    oldy = y;
                }
            }
            plot(col, row, (color % 16));
            plot(col, 240-row, (color % 16));
        }
    }
}
```

(continued)

```
/*
┌─────────────────────────────────────────────────────┐
│                                                     │
│   move_cursor() = Moves the cursor on the screen    │
│                                                     │
└─────────────────────────────────────────────────────┘
*/

Julia move_cursor(void)
{
      int i,ch=0, col, row, min_x = 3, min_y=3,
            max_x=317, max_y=237;
      static int c_col=160, c_row=120, flag=0;
      float r, dx, dy, theta, x, y;
      char zero_buf[40];
      Julia c;

      memset(zero_buf,0x00,40);
      zero_buf[0] = 0x80;
      zero_buf[39] = 0x03;
      c.cx = ((float)c_col*6.8/xres) - 2.2;
      c.cy = -(((float)c_row*5.2/yres) - 1.3);
      gotoXY(3,20);
      printf("Arrows move cursor on Mandelbrot Set");
      gotoXY(3,22);
      printf("Hit 'Enter' for final Julia Set...");
      gotoXY(3,24);
      printf("c: (%f,%f)       ",c.cx, c.cy);
      if (flag == 0)
      {
            arrow(c_col, c_row);
            flag = 1;
      }
      while (ch != ENTER)
      {
            x = 0;
            y = 0;
            i = 0;
            while (!kbhit())
            {
                  dx = x - c.cx;
                  dy = y - c.cy;
```

(continued)

```
        if (dx > 0)
                theta = atan(dy/dx)*0.5;
        else
        {
                if (dx<0)
                        theta = (3.14159 +
                                atan(dy/dx))*0.5;
                else
                        theta = 0.78539;
        }
        r = sqrt(sqrt(dx*dx + dy*dy));
        if (rand() < 16384)
                r = -r;
        x = r*cos(theta);
        y = r*sin(theta);
        col = (x + 2.2)*xres/8.8 + 320;
        row = 240 - ((y + 1.4)*yres/5.6);
        if ((i>20) && (col<638) && (col>321) &&
                (row>3) && (row<238))
                plot(col, row, 10);
        i++;
}
ch = bioskey(0);
arrow(c_col,c_row);
switch(ch)
{
        case UP_ARROW:
                if (c_row>min_y)
                        c_row--;
                break;
        case SHIFT_UP_ARROW:
                if (c_row>(min_y+4))
                        c_row -= 5;
                break;
                case DOWN_ARROW:
                        if (c_row<max_y)
                                c_row++;
                        break;
                case SHIFT_DOWN_ARROW:
                        if (c_row<(max_y-4))
                                c_row += 5;
```

(continued)

```
                                break;
                        case LEFT_ARROW:
                                if (c_col>min_x)
                                        c_col--;
                                break;
                        case SHIFT_LEFT_ARROW:
                                if (c_col>(min_x+4))
                                        c_col -= 5;
                                break;
                        case RIGHT_ARROW:
                                if (c_col<max_x)
                                        c_col++;
                                break;
                        case SHIFT_RIGHT_ARROW:
                                if (c_col<(max_x-4))
                                        c_col += 5;
                }
                arrow(c_col,c_row);
                c.cx = ((float)c_col*6.8/xres) - 2.2;
                c.cy = -(((float)c_row*5.2/yres) - 1.3);
                if (ch != ENTER)
                        for(i=2; i<239; i++)
                                movedata(FP_SEG(zero_buf),
                                        FP_OFF(zero_buf),
                                        0xA000, i*80 + 40,40);
                gotoXY(3,24);
                printf("c: (%f,%f)      ",c.cx, c.cy);
        }
        return c;
}

/*
╔══════════════════════════════════════════════════╗
║                                                  ║
║        arrow() = Plots an arrow on the screen    ║
║                                                  ║
╚══════════════════════════════════════════════════╝
*/

void arrow (int x, int y)
{
        int i,j;
```

(continued)

```
      for (i=0; i<8; i++)
            for (j=0; j<=i; j++)
                  plotxor(x+j,y+i);
      for (i=0; i<5; i++)
            plotxor(x+i,y+8);
      plotxor(x,y+9);
      plotxor(x+2,y+9);
      plotxor(x+3,y+9);
      plotxor(x+4,y+9);
      plotxor(x,y+10);
      plotxor(x+4,y+10);
      plotxor(x+5,y+10);
      plotxor(x+4,y+11);
      plotxor(x+5,y+11);
      plotxor(x+5,y+12);
      plotxor(x+6,y+12);
      plotxor(x+5,y+13);
      plotxor(x+6,y+13);
      plotxor(x+6,y+14);
      plotxor(x+7,y+14);
      plotxor(x+6,y+15);
      plotxor(x+7,y+15);
}

/*

            julia() = Generates the Julia set

*/

void julia(long int p, long int q)
{
      int i, col_start, col_end;
      long int Pmax, Pmin, Qmax, Qmin;
      long int Q[240], P, deltaP, deltaQ, x, y, xt, yt,
            xsq, ysq;
      float temp;

      Pmax = 2.2 * SCALE;
      Pmin = -2.2 * SCALE;
      Qmax = 1.4 * SCALE;
```

(continued)

```
        Qmin = -1.4 * SCALE;
        deltaP = (Pmax - Pmin)/(xres/2);
        deltaQ = (Qmax - Qmin)/(yres/2);
        for (row=0; row<(yres/2); row++)
              Q[row] = Qmax - row*deltaQ;
        col_start = 321;
        col_end = xres-3;
        for (col=col_start; col<col_end; col++)
        {
              for (row=241; row<478; row++)
              {
                    x = Pmin + (col - col_start) * deltaP;
                    y = Q[row-241];
                    xsq = ysq = 0;
                    for (color=0; ((color<64) && (xsq + ysq
                          < MAXSIZE)); color++)
                    {
                          xt = (x>>11);
                          yt = (y>>11);
                          xsq = xt*xt;
                          ysq = yt*yt;
                          x = xsq - ysq + p;
                          y = 2*xt*yt + q;
                    }
                    if (color >= 64)
                    {
                          temp = (double)(xsq +
                                ysq)/(double)SCALE;
                          color = ((int)(temp*9.0))%6 + 9;
                    }
                    else
                          color = 1;
                    plot(col, row, (color % 16));
              }
        }
  }
```

(continued)

```
/*
╔══════════════════════════════════════════════════════════╗
║                                                          ║
║              setmode() = Sets video mode                 ║
║                                                          ║
╚══════════════════════════════════════════════════════════╝
*/

void setmode(int mode)
{
      reg.x.ax = mode;
      int86 (0x10,&reg,&reg);
}

/*
╔══════════════════════════════════════════════════════════╗
║                                                          ║
║              cls() = Clears the screen                   ║
║                                                          ║
╚══════════════════════════════════════════════════════════╝
*/

void cls(int color)
{
        reg.x.ax = 0x0600;
        reg.x.cx = 0;
        reg.x.dx = 0x1E4F;
        reg.h.bh = color;
        int86(0x10,&reg,&reg);
}

/*
╔══════════════════════════════════════════════════════════╗
║                                                          ║
║           gotoXY() = Sets cursor position                ║
║                                                          ║
╚══════════════════════════════════════════════════════════╝
*/

void gotoXY(int col, int row)
{
      reg.h.ah = 0x02;
      reg.h.bh = 0;
      reg.h.dh = row - 1;
```

(continued)

```
        reg.h.dl = col - 1;
        int86(0x10,&reg,&reg);
}

/*
```

plot() = Plots a point on the screen at a designated position using a selected color for 16 color modes.

```
*/

void plot(int x, int y, int color)
{
        int dummy,mask;
        char far * address;

        address = (char far *) 0xA0000000L + (long)y *
              xres/8L + ((long)x / 8L);
        mask = 0x8000 >> (x % 8);
        outport(0x3CE,mask | 8);
        outport(0x3CE,0x0205);
        dummy = *address;
        *address = color;
        outport(0x3CE,0x0005);
        outport(0x3CE,0xFF08);
}

/*
```

plotxr() = Plots a point on the screen at a designated position XORing with existing data for 16 modes.

```
*/

void plotxor(int x, int y)
{
        int dummy, mask;
```

(continued)

```
    char far * address;

    address = (char far *) 0xA0000000L + (long)y *
          xres/8L + ((long)x / 8L);
    mask = 0x80 >> (x % 8);
    outport(0x3CE, (mask << 8) | 0x08);
    outport(0x3CE, 0x1803);
    outport(0x3C4,0xFF02);
    dummy = *address;
    *address = 0;
    outport(0x3C4,0x0F02);
    *address = 0xFF;
    outport(0x3C4,0xFF02);
    outport(0x3CE,0x03);
    outport(0x3CE,0xFF08);
}
```

12

Using Pointers in C

One of the most difficult things to understand in working with the C language is the use of pointers. It takes a lot of experience to be able to keep track of what is a pointer and what isn't and to use the proper nomenclature so that all of your statements work out correctly. However pointers are essential if you are to take advantage of the full capability of C. They provide an elegant mechanism for doing all kinds of advanced argument manipulation. They also overcome the limitation of C of only being able to return a single value from a function. We've already briefly touched on the subject of pointers in Chapter 3. In this chapter we'll look in detail at just what a pointer is and how it is used.

What Is a Pointer?

Every argument in C may be referred to in two ways. One way is to refer to the actual value of the argument. This value is kept in a memory location somewhere. The second way to refer to the argument is by a pointer, which contains the address of the memory location where the value is stored. When you use the pointer, you know where the argument is, but not what its contents may be. The problem is that when you see a name in the body of a program, you do not know for sure whether it is the name of an argument or the name of a pointer. To find out which, you need to go back to the place in the program where the name is defined. First, let's look at a few name definitions

```
int a, b, c;
long d, e, f;
float x, y, z;
double u, v, w;
```

All these names, *a, b, c, d, e, f, u, v, w, x, y*, and *z*, are the names of arguments. If you tell C to do something with *a*, for example, it will take the value of the argument *a* and use it according to your instructions. Now, suppose that for some reason you want to work with a pointer to *a*. You then use the expression *&a*. This means that you are working with the address at which *a* is stored. This is no problem when *a* has been defined as an argument, as we just did previously. However, it is also possible to define *a* as a pointer to the address of an argument, as we shall see in just a minute. In such a case, the rules are changed. We need to refer to **a* if we want the value of the argument pointed to by *a*. (The asterisk is also used to denote multiplication, which makes things more difficult, since we have to decide the meaning of the asterisk from the context of the expression in which it occurs.) As an added complication, if we use the expression *&a* and *a* is already defined as a pointer, then we are using a pointer to a pointer to an argument value. As you can see, this begins to get complicated pretty quickly. We hope you'll never need to use addresses of addresses, etc., but if you do, C allows you to do this.

Suppose that we have some names defined like this

```
int *a, *b;
float *c, *d;
char *e;
```

The asterisk preceding the name means that the name is the name of a pointer. (This is a third meaning of the asterisk, which only applies when we are defining the type of an argument.) While we have specified that the pointer is the address of an argument of a particular type, the way that memory is allocated is totally different from when we define an argument. The statement *int i;* says that we have defined an integer named *i* and also indicates that space in computer memory has been allocated in which to store this value. The statement *int *i;* says that we have defined a pointer to the address of an integer and that space in computer memory has been allocated for this pointer. The amount of space allocated is whatever is required to hold the argument address; this size depends on the memory model used and has no relation to the type of data. Since only the location of the address information is defined, you have not so far allocated any space for the value of the argument. You have to do this by a separate action; otherwise an attempt to enter a value for the argument results in overlapping some already used part of memory, often with disastrous results. There are two ways of assigning memory to values when only a pointer is defined. The first works like this

```
int a, *b;

b = &a;
```

We have defined *a* as an integer argument (and also allocated memory space for it). We have defined *b* as a pointer giving the address of an integer. In the lower statement, we have set *b* equal to the address of *a*. From now on, we can refer to the address of *a* by *b*. At this point, *b* is just the same as *&a*.

The other method for allocating space to store values of arguments that are only defined by their address pointers is through the use of the *malloc* and *cmalloc* functions, which actually allocate a block of memory of a specified size for you to use to store the values pointed to by your pointer. How to use these functions will be described in Chapter 17.

Now let's look at some arrays:

```
char g[40];
float k[22];
int i[80];
```

Arrays in C are defined in such a way that *g* is a pointer to the address of an array of 40 values of type *char*. Now, if you use the name *g* you are referring to a pointer that tells you where this array starts, but if you use *g[0]*, *g[5]*, or *g[24]*, you are referring to the value of that member of the array, not to a pointer to its address. Therefore *g* is the same as *&g[0]*.

Next, let's look at a couple of multidimensional arrays, where the situation gets much worse:

```
int m[6][8];
float pk[4][10];
```

Suppose you refer to *m*. This is a pointer to the address of the beginning of the array *m*. Now suppose you refer to *m[1]*. This still doesn't identify any particular member of the array *m*. Therefore *m[1]* is a pointer to the address of the beginning of the array section *m[1]*.

If you refer to *m[1][2]*, however, you have finally identified a particular member of the array. In this case you are alluding to the value of that particular argument *m[1][2]*. This is the eleventh member of the *m* array, since we have eight members of the array in the *m[0]* row (as specified by the second or right-hand member of the array definition) and then the 0 and 1 members of the *m[1]* row before we come to *m[1][2]*. To generalize, if you specify the array name followed by a number in brackets for each number in brackets that is given in the array definition, then you have enough information to identify a specific member of the array, in which case your name is an argument; whereas if you don't have enough numbers in brackets to identify an array member completely, your name will be taken as a pointer to a location at the beginning of or within the array. If you define a pointer by *m[1]*, for example, this refers to the address of the second row of the array. (There is no way that you can go the other way and look at columns of the array.) Some compilers will not allow you to pass the pointer *m* for a multidimensional array when a pointer is expected. Instead, to make the compiler function properly, you will have to pass *m[0]*.

Now, we're going to look at a little program that makes use of some of these concepts:

```
#include <stdio.h>
#include <math.h>

void subr(int x);
void sub2(int *x);

void main(void)
{
      int a;
      int b[20];
      int c[3][4];
      int *d;

      clrscr();
      printf("\naddress of b: %p",b);
      printf("\naddress of c: %p",c);
      printf("\naddress of c[0]: %p",c[0]);
      printf("\naddress of c[1]: %p",c[1]);
      a = 5;
      b[0] = 6;
      d = c[1];
      c[0][0] = 3;
      c[0][1] = 4;
      c[0][2] = 7;
```

(continued)

```
        c[1][0] = 8;
        printf("\naddress given by d: %p",d);
        printf("\na = %d",a);
        printf("\nb[0] = %d",b[0]);
        printf("\nc[0][0] = %d",c[0][0]);
        subr(*b);
        subr(*c[0]);
        subr(*c[1]);
        subr[*d];
        sub2(b);
        sub2(c);
        sub2(c[1]);
        sub2(d);
        getch();
}

void subr(int x)
{
        printf("\nx = %d",x);
}

void sub2(int *y)
{
        printf("\nvalue pointed to by y = %d",y[0]);
}
```

We have two functions. The first, *subr*, passes an integer *x*. It displays the value of this integer. The second function, *sub2*, passes a pointer *y*, which is the address of an argument. It displays the value of the argument pointed to by *y*. Now look at the main program, which begins by clearing the screen. It then displays the addresses of *b, c, c[0]*, and *c[1]* and later *d*. The data type printed out by each *printf* statement is of type *p*, which is an address pointer, and the argument referenced by each *printf* statement is, as it should be, a pointer. (The argument *b* is a pointer to the beginning of the array *b*. The arguments *c* and *c[0]* both are pointers to the array c. The argument *c[0]* is a pointer since *c* is a multidimensional array.) The argument *c[1]* is a pointer to the beginning of the second row of the *c* array. The argument *d* is a pointer that has been set equal to the pointer *c[1]*. The various arguments and array members are set to selected values and then *printf* statements are used to display the single variable and the first member of each array. Next, the values of three array members are passed to *subr* for display. Note what happens when the array members are not fully identified. When **b* is passed, the value that is passed is that of *b[0]*. When **c[0]* is passed, the value that is passed is that of *c[0][0]*. When **c[1]* is passed, the value that is passed is that of *c[1][0]*. Next, four pointers are passed to *sub2* for display. When *b* is passed, the *printf* statement in the function displays *b[0]*.

When *c* is passed, the array member that is actually displayed is *c[0][0]*. When *d* is passed the array member that is actually displayed is *c[1][0]*. The resulting display created by this program is:

```
address of b: FFCC
address of c: FFB4
address of c[0]: FFB4
address of c[1]: FFBC
address given by d: FFBC
a = 5
b[0] = 6
c[0][0] = 3
x = 6
x = 3
x = 8;
x = 8;
value pointed to by y = 6
value pointed to by y = 3
value pointed to by y = 8
value pointed to by y = 8
```

You should now be getting a feel for how pointers work. Here's another simple example that shows a practical use of pointers.

```
#include <stdio.h>

void squares(int *x, int *y)

void main(void)
{
      int a, b;

      a = 4;
      b = 6;
      squares(&a,&b);
      printf("\na square = %d   b square = %d",a,b);
}

void squares(int *x, int *y)
{
      *x = *x * *x;
      *y = *y * *y;
}
```

This is a good example because it demonstrates a lot of the things about pointers that tend to be confusing. What we want to do in this program is have a function that returns the squares of two arguments that are passed to it. However, a C function can only return one argument. If we simply passed *a* and *b* to the *squares* function, *squares* would be sent copies of the values of *a* and *b*. We could square them within the function, but would have no way of getting the new values back to the main program. So what we do is define the two arguments that are passed to *squares* to be pointers to the arguments. Thus when we define an argument in the prototype and in the function definition as *int *x* it means that what is being passed to the function is a pointer to the location where the actual value is found. Now in the actual call of the function we have *squares(&a, &b);*. What you have to keep straight is that in the function definition the *int *x* means that the type of data being passed is a pointer to a function and that in the function call, the *&a* defines the actual data passed to be a pointer to the argument *a*. This makes the actual data consistent in its type with the type of data defined in the function definition so that we have no problems. If there were an inconsistency, the compiler would report this as an error when it attempts to compile the program. If you're new at this, it's a little hard to get used to the fact that we use the * in the function definition and the & sign in the actual function call to achieve the proper consistency, but after a while, this will become second nature to you.

Now let's look at the function itself. We have two pointers passed to it; we want to do some work with the values themselves. Again we use the asterisk, where in this context it means to operate on the value found in the pointer location. So we have **x = *x * *x;*. The asterisk on the left-hand side of the equation means that we are going to put our result in the location pointed to by *x*. The first and third asterisks on the right-hand side mean that we are going to operate upon two contents of the value that is pointed to by *x*. The middle asterisk means that we are going to perform a multiplication. Thus we square the original value and put it back where the value was originally found. This location has been defined as the location of *a* in the main program, so that when we go back to the main program we'll find the squared result at the location pointed to by *a*. Similar reasoning applies to the line that squares the value pointed to by *y* and returns it to the location, which is pointed to both by *y* in the function and by *b* in the main program. Thus we have squared the two numbers and gotten the values back to the main program. At first glance the two asterisks side by side in the two lines of code in the function look pretty silly and it takes a second look to realize that the first means multiplication and the second means to operate on the contents of the pointer. However, the compiler has no problem understanding which is which and operating on the expression correctly.

The Plasma Fractal Display

The *plasma* fractal program produces a very interesting varicolored display that has a cloudlike appearance on a VGA. Once the initial display is generated, the 256 colors of the mode 19 graphics display can be cycled to give a display of constantly changing colors that is almost hypnotic. The program is included in this chapter because we are using a very interesting application of pointers to do the color cycling. The program is listed in Figure 12-1 and a typical display appears in Plate 10.

The main program begins by setting the video graphics mode, initializing the 256-color palette, and then calling *randomize*, which initializes the random number generator with a random number that is different each time the program is run. The program then plots a point of random color at each corner of the display. Next, it calls *subdivide*, which selects colors and plots points for all of the rest of the display. The program then pauses until a key is struck. It then calls *rotate_colors*, which rotates the colors and thereby produces a dynamic display. When *rotate_colors* ends, the program waits for a keystroke and then resets the display mode to the standard text mode and terminates.

Initializing the Palette

In working with the 256-color palette, we make use of two functions that call the ROM BIOS video services to send new color information to the color registers. The first of these is *setVGAreg*. This function sets register *a* to 0x1012 to call video service 10, subservice 10. This service sets the color register number specified to the red, green, and blue values that are passed by the function. Thus it changes the color of only one register. The second function used is *setVGApalette*. This function can transfer color data from a data array to a number of color registers in one operation. The parameter passed to this function is the address of the buffer containing the palette data. The first color register to be affected is listed in register *bx* and the number of registers whose data is to be changed is specified in register *cx*. As we modify color registers, either in initializing or in cycling colors, we are always going to keep color register 0 set to black, so we have set up the *setVGApalette* function to start with color register 1 and write data to 255 registers so that all color registers are changed by this function except register 0. The *initPalette* function sets up the initial values for the color data array *palette*. For reasons that will become apparent later, we have an array of 512 sets of three colors. Each color set occurs twice in this array, once within the first 255 sets of three colors and 255 sets later. A single *for* loop is used for the initialization. The first six statements in this *for* loop apply to the initial group of 85 shades in which the red component is 0 and the green and blue are shaded throughout their range. The next six statements initialize a second group of 85 shades in which the blue component is held to 0. The final six statements initialize a third group of 85 shades in which the green component is held to 0. When all this data has been set up in the color data array *palette* the *setVGAreg* function is called to set the 0th color register to black (0,0,0). The *setVGApalette* function is then called to set color registers 1 to 255 from the color data array *palette*.

The *subdivide* Function

The *subdivide* function is passed four points which mark the corners of a rectangle on the display. It first finds the coordinates (x and y), which denote the center of this rectangle. If the x coordinate of this midpoint corresponds to the x value of the top left corner of the rectangle, then all of the pixels on the display have been colored so the function needs to do nothing further. It therefore returns to the calling function. Otherwise, for each of the the four points that are the midpoints of the four sides of the rectangle, the function finds a random number whose + and - limits are the distance between the endpoints of that side. It then adds this random value to the average of the colors of the two endpoints. Thus, the farther separated the endpoints are, the more the random color component can cause the colors to vary. For points very close together, there can be little color change. Each of these four points is plotted to the screen only if that pixel is not already colored. The function next plots the midpoint of the rectangle to the screen in a color that is the average of the colors of the four corner points of the rectangle. Finally, the function recursively calls itself to process each of the four rectangles defined by the midpoint of the rectangle, the midpoint of each of the four sides, and the rectangle corner point of that quadrant. This is a good example of recursion. We won't discuss it further here, since it is explained fully in Chapter 19. When you read that chapter, come back and look again at this program.

Rotating Colors

This function keeps rotating the palette colors to give interesting effects on the screen. It begins with an infinite *for* loop, one which we can only escape from with a *break* statement. At each iteration of the *for* loop, the function first calls *kbhit* to see if a key has been struck. If so, the key is read, and if it was the *Ent* key, the *break* is called and the function terminates. If the key was a number from 1 to 9, the parameter *last_step* is set to 4 times the number selected and the function proceeds. For other keys, the function just ignores them and continues. What we want to do is move the contents of each color register down to the next lower register. (The contents of color register 0 is dropped completely.) Then we place a new set of color values in the last color register. We select a new set of three color values by taking three random numbers from 0 to 63. How we get from the old last color to the new one determines what the display will look like; fast and slow changes make completely different looking displays. Basically what we do is enter a *for* loop that iterates for as many times as the setting of the parameter *last_step*. At each iteration, we add to the old color setting *1/last_step* of the difference between the old and new color values. Then we rotate down the color registers and place the computed value in the last register. As you can see, the higher the value of *last_step*, the closer the color in the last register will be to that in the next to last register for each iteration, and therefore the slower the color changes will appear to move on the screen.

We have all of the color register data in the *palette* array and could use a *for* loop to move each set of three array values down to the next lower array component. Then we would just store the computed next set of values in the last set of array members. This requires many operations and slows things up considerably. What we are going to do instead is a little trickery using pointers. Instead of shifting every set of colors in the *palette* array down one, when we call *setVGApalette* to use the *palette* data to set the color registers, we are going to give it an address pointer to *palette* that has moved up one set of color values. We do this by incrementing a parameter *step*, which is part of the pointer designator at each pass through the color cycling *for* loop. This is fairly simple, but there are a couple of problems that must be solved in order for everything to work properly. First, as we move up the array with our starting pointer, there will be data at the end that must be the same as the beginning data to get all the color registers set up properly. This is why we defined the array to have 512 sets of data even though there are only 256 color registers and this is why in *initPalette* we loaded a duplicate set of color data above the first. Of course we can't keep increasing *step* forever, so we have a final line in the *for* loop that resets *step* to 1 once it reaches 255. The final thing that we have to look out for is that when we are inserting new color data into the last register, we have to make sure that it goes into the set of color values in *palette* that will be sent to the last color register and that the duplicate set of the same colors will also be inserted into *palette* in the proper place. It's not easy to figure this out in your head when you are writing such a program, but with a little trial and error you can usually get it all straightened out. If you look at the listing for this program, you can see what the right values are for this case and use it as a model for your own.

There is one final point to discuss. You'll note a couple of unfamiliar lines involving *while* loops just before we set the VGA palette. These are included because if we set the palette registers while we are in the middle of writing a display screen, there will be unwanted transients that mar the display. Therefore we use these two loops to determine the beginning of a vertical retrace so that we can change the color registers when the screen is blank. Each loop ends with a semicolon so that the loop repeats until the condition is met but doesn't do anything else. The first loop checks for the vertical retrace bit to be nonzero, which means that vertical retrace is taking place. This assures that we don't reach this point somewhere near the end of vertical retrace and attempt to change color registers when there is not enough time. The loop cycles until vertical retrace is over. Then the next *while* loop takes over and cycles until the next vertical retrace just begins. At this point, we change the color registers.

This is just one example of the clever use of C pointer capabilities to save a lot of computation time. As you progress in working with pointers, you'll begin to discover other clever ways in which to make use of them. It's possible to get pretty subtle with pointers, however. If you think that someone else will have to maintain your software in the future, don't get so cute with your pointer use that no one else can ever figure out what you did.

Figure 12-1. Listing of Plasma Program

```
/*
 ┌──────────────────────────────────────────────────────────┐
 │                                                          │
 │        PLASMA = Creates a plasma display an a VGA        │
 │                                                          │
 │              By Roger T. Stevens   7-26-92               │
 │                                                          │
 └──────────────────────────────────────────────────────────┘
*/

#include <dos.h>
#include <stdio.h>
#include <stdlib.h>
#include <time.h>
#include <conio.h>

void initPalette(void);
void plot(int x, int y, int color);
int readPixel(int x, int y);
void rotate_colors(void);
void setmode(int mode);
void setVGApalette(unsigned char *buffer);
void setVGAreg(int reg_no, int red, int green, int blue);
void subdivide(int x1, int y1, int x2, int y2);


union REGS reg;
struct SREGS inreg;

int xres=319, yres=199;
unsigned char palette[512][3];
char ch;
int i,j,k;

void main()
{
	setmode(0x13);
	initPalette();
	randomize();
	plot(0,0,random(255) + 1);
	plot(xres,0,random(255) + 1);
	plot(xres,yres,random(255) + 1);
	plot(0,yres,random(255) + 1);
	subdivide(0,0,xres,yres);
	ch = getch();
	rotate_colors();
	getch();
	setmode(0x03);
}
```

(continued)

```
/*
    initPalette() = Sets the colors of the VGA palette
*/

void initPalette(void)
{
      int max_color = 63;
      int index;

      for (index=0; index<85; index++)
      {
            palette[index][0] = 0;
            palette[index][1] = (index*max_color) / 85;
            palette[index][2] = ((86 - index)*max_color) /
                  85;
            palette[index+255][0] = 0;
            palette[index+255][1] = (index*max_color) /
                  85;
            palette[index+255][2] = ((86 - index) *
                  max_color) / 85;
            palette[index+85][0] = (index*max_color) / 85;
            palette[index+85][1] = ((86 - index) *
                  max_color) / 85;
            palette[index+85][2] = 0;
            palette[index+339][0] = (index*max_color) /
                  85;
            palette[index+339][1] = ((86 - index) *
                  max_color) / 85;
            palette[index+339][2] = 0;
            palette[index+170][0] = ((86 - index) *
                  max_color) / 85;
            palette[index+170][1] = 0;
            palette[index+170][2] = (index*max_color) /
                  85;
            palette[index+424][0] = ((86 - index) *
                  max_color) / 85;
            palette[index+424][1] = 0;
            palette[index+424][2] = (index*max_color) /
                  85;
      }
      setVGAreg(0,0,0,0);
      setVGApalette(palette[0]);
}
```

(continued)

```
/*
 ┌──────────────────────────────────────────────────────────┐
 │                                                          │
 │          plot() = Plots a point on 256 color screen      │
 │                                                          │
 └──────────────────────────────────────────────────────────┘
*/

void plot(int x, int y, int color)
{
      char far *address;
      address = (char far *)(0xA0000000L + 320L * y + x);
      *address = color;
}

/*
 ┌──────────────────────────────────────────────────────────┐
 │                                                          │
 │        readPixel() = Reads a pixel from the screen       │
 │                                                          │
 └──────────────────────────────────────────────────────────┘
*/

int readPixel(int x, int y)
{
      reg.h.ah = 0x0D;
      reg.x.cx = x;
      reg.x.dx = y;
      int86 (0x10,&reg,&reg);
      return (reg.h.al);
}

/*
 ┌──────────────────────────────────────────────────────────┐
 │                                                          │
 │         rotate_colors() = Rotates the VGA colors         │
 │                                                          │
 └──────────────────────────────────────────────────────────┘
*/

void rotate_colors(void)
{
      int i, j, k, old_red, old_green, old_blue, new_red,
            new_green, new_blue, last_step=32, step = 1;
      char ch;

      for(;;)
      {
            if (kbhit() != 0)
            {
                  ch = getch();
                  if (ch == 0x0D)
                        break;
                  else
                  {
```

(continued)

```
                    if ((ch - '0' <= 9) && (ch - '0' >
                         0))
                         last_step = 4* (int)(ch -
                              '0');
               }
          }
          old_red = palette[step + 253][0];
          old_green = palette[step + 253][1];
          old_blue = palette[step + 253][2];
          new_red = rand() % 64;
          new_green = rand() % 64;
          new_blue = rand() % 64;
          for (j=1; j<last_step; j++)
          {
               palette[step + 254][0] = old_red +
                    ((new_red-old_red)*j)/last_step;
               palette[step-1][0] =
                    palette[step+254][0];
               palette[step + 254][1] = old_green +
                    ((new_green-old_green) *
                    j)/last_step;
               palette[step-1][1] = palette[step +
                    254][1];
               palette[step + 254][2] = old_blue +
                    ((new_blue-old_blue)*j) /
                    last_step;
               palette[step-1][2] = palette[step +
                    254][2];
               while ((inportb(0x3DA) & 0x08) != 0);
               while ((inportb(0x3DA) & 0x08) == 0);
               setVGApalette(palette[step]);
               step = step%255 + 1;
          }
     }
}

/*

                    setmode() = Sets video mode

*/

void setmode(int mode)
{
     reg.x.ax = mode;
     int86 (0x10,&reg,&reg);
}
```

(continued)

```
/*
    setVGApalette() = Sets all VGA registers except register 0
*/

void setVGApalette(unsigned char *buffer)
{
        reg.x.ax = 0x1012;
        segread(&inreg);
        inreg.es = inreg.ds;
        reg.x.bx = 1;
        reg.x.cx = 255;
        reg.x.dx = (int)&buffer[0];
        int86x(0x10,&reg,&reg,&inreg);
}

/*
        setVGAreg() = Sets an individual VGA color register
*/

void setVGAreg(int reg_no, int red, int green, int blue)
{
        reg.x.ax = 0x1010;
        reg.x.bx = reg_no;
        reg.h.ch = red;
        reg.h.cl = green;
        reg.h.dh = blue;
        int86(0x10,&reg,&reg);
}

/*
    subdivide() = Divides up a display section and fills with
                              color.
*/

void subdivide(int x1, int y1, int x2, int y2)
{
        int x, y, color, dist;

        x = (x1 + x2) >> 1;
        y = (y1 + y2) >> 1;
        if (x == x1)
                return;
        dist = x2 - x1;
```

(continued)

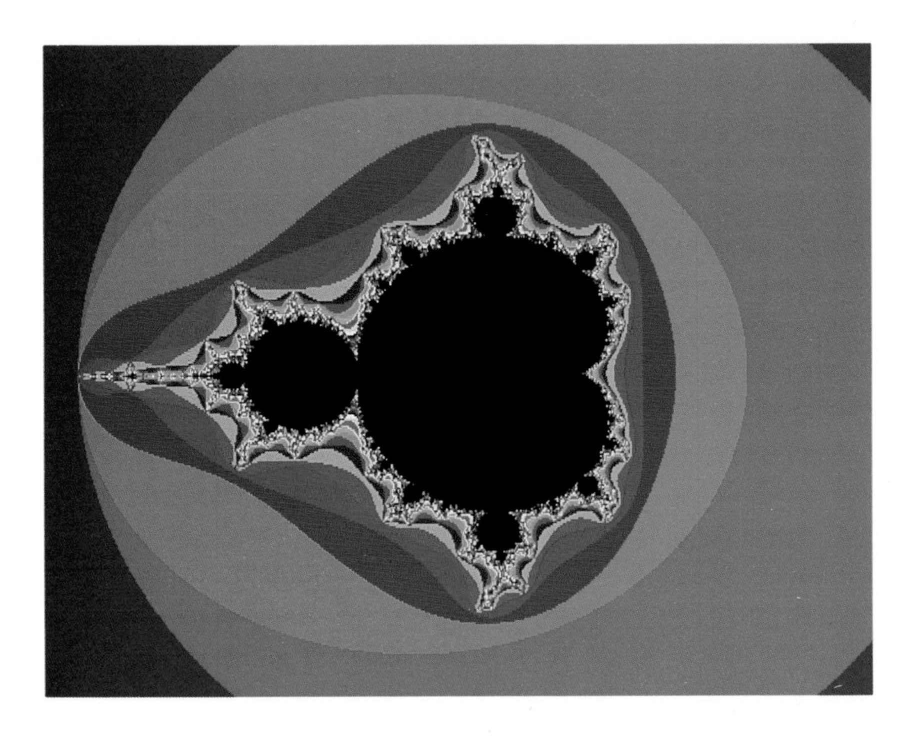

Plate 1. Mandlebrot Set

Plate 2. Hyperbolic Cosine Fractal

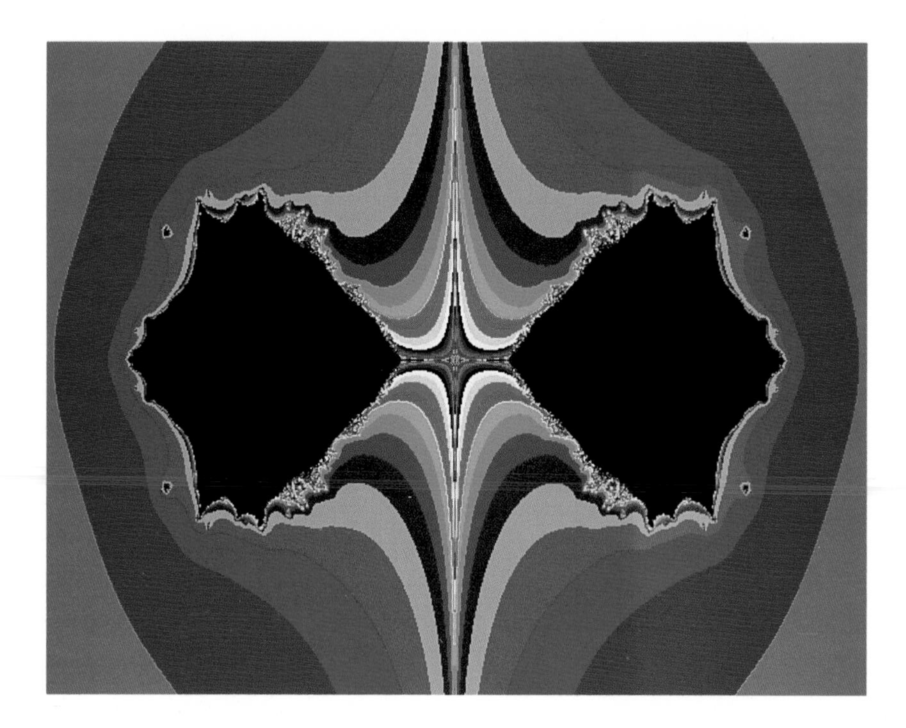

Plate 3. Legendre Polynomial Fractal

Plate 4. Julia Set

Plate 5. Dragon Curve

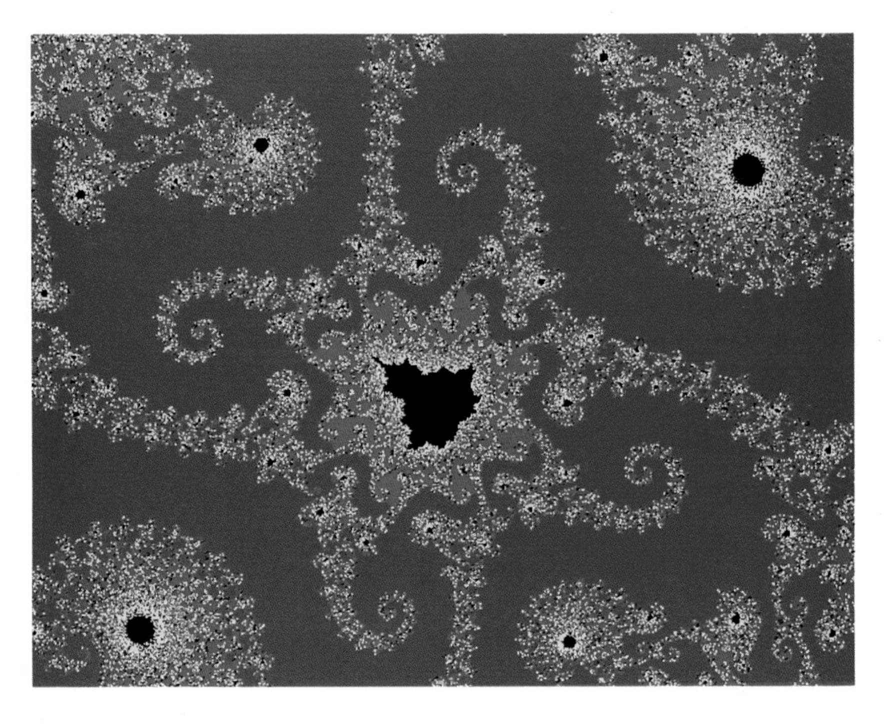

Plate 6. Expanded Mandlebrot Set with Specified Colors

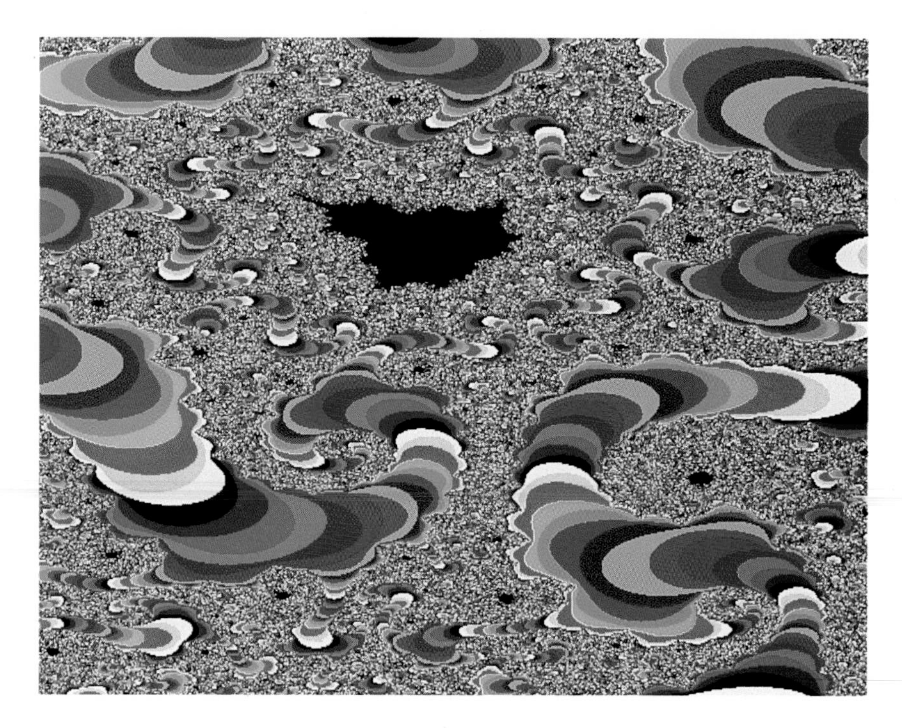

Plate 7. Mandlebrot Set with Selectable Degree of Expansion

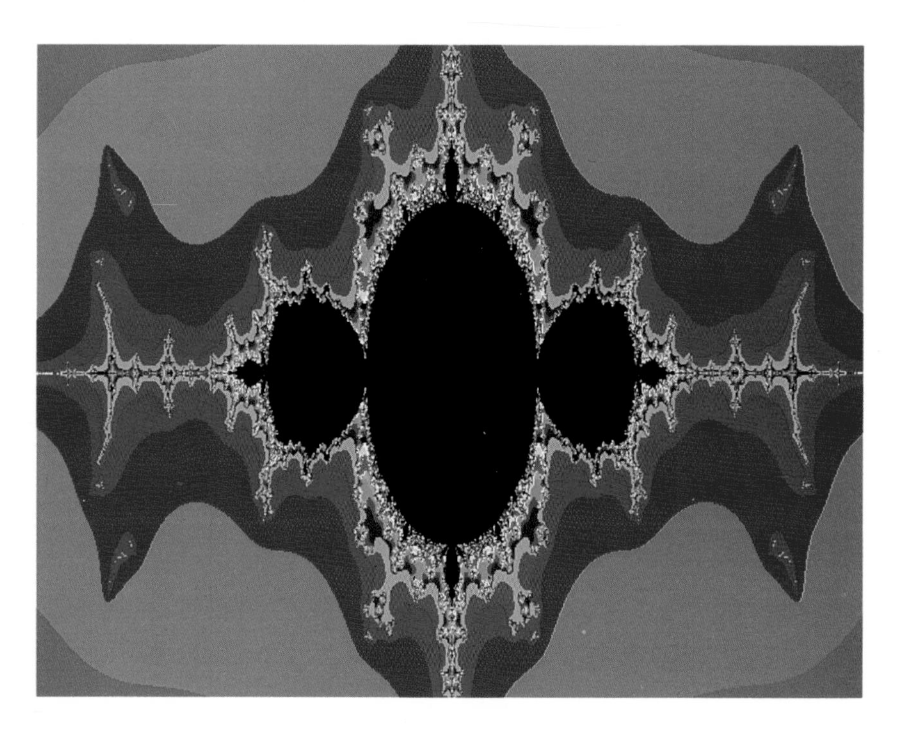

Plate 8. The Tchebychev C_5 Fractal Curve

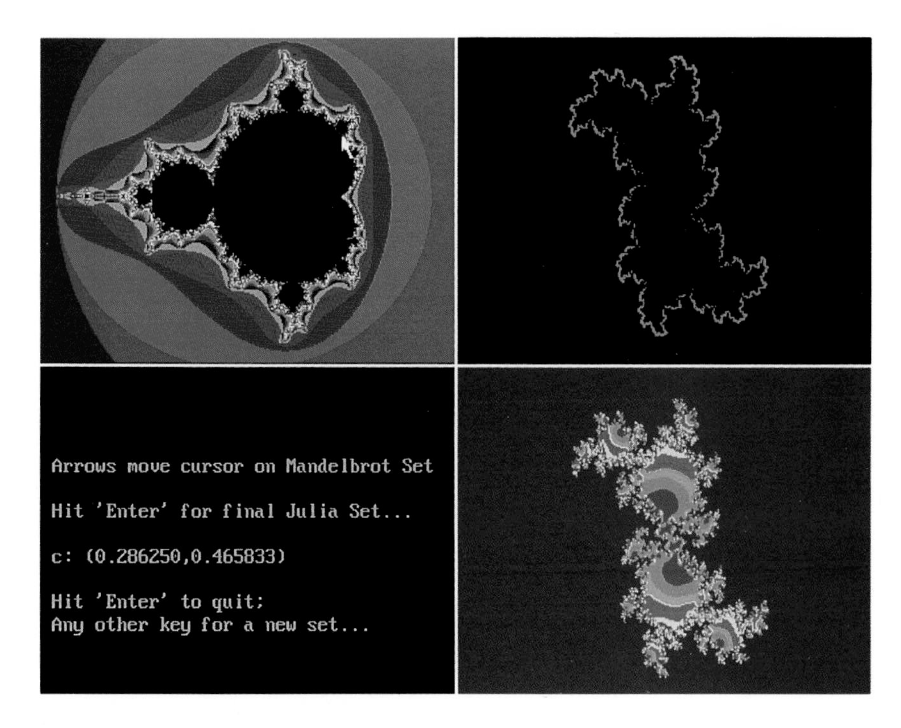

Plate 9. The Mandlebrot Set as a Map of Julia Sets

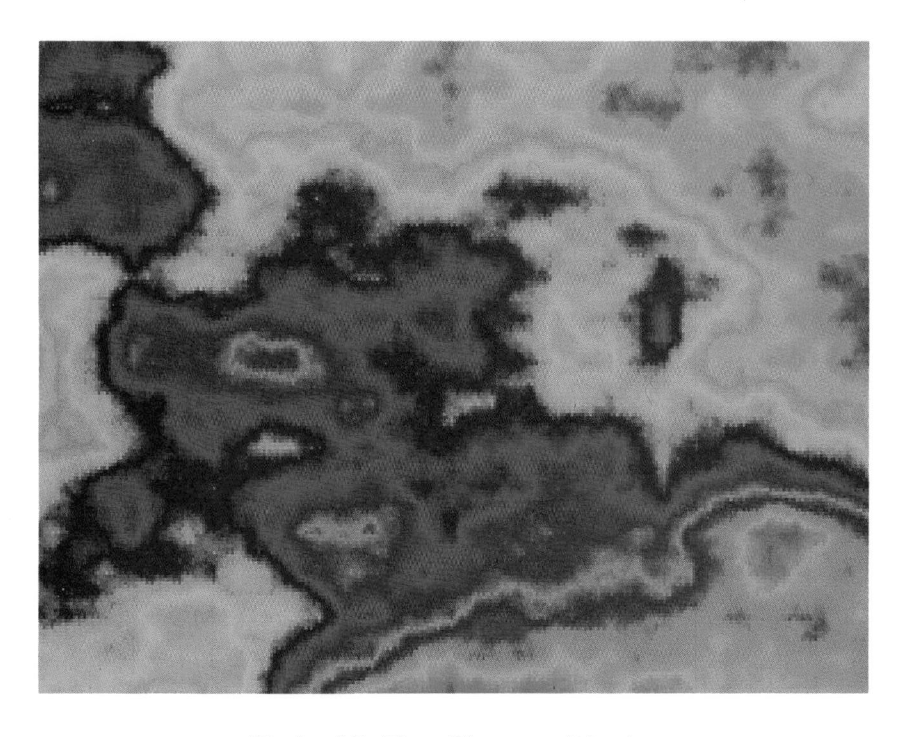

Plate 10. The Plasma Display

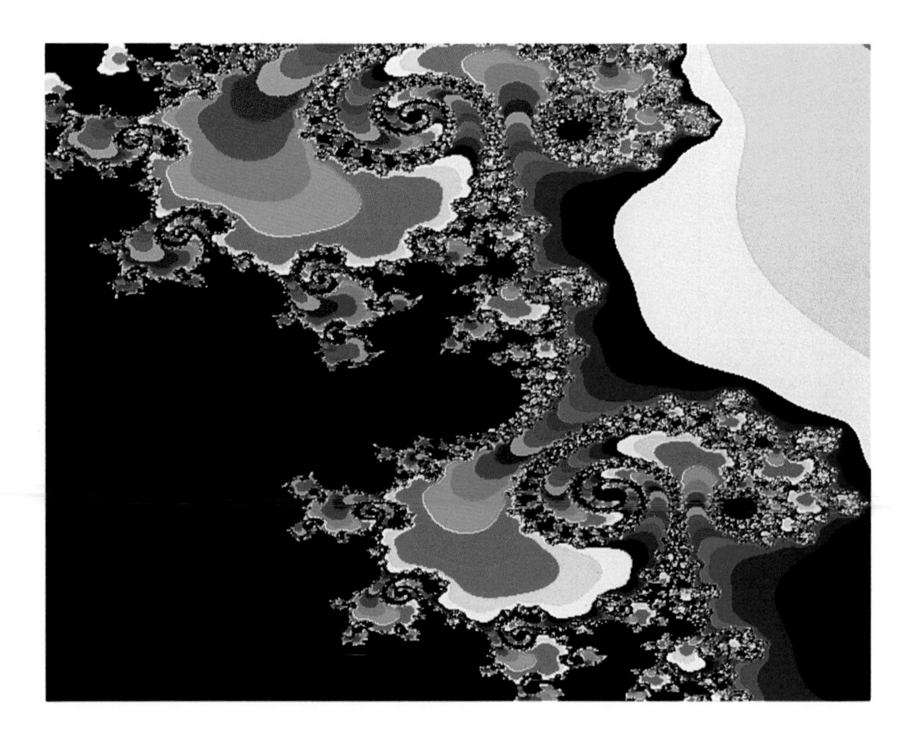

Plate 11. Expanded Tchebychev C_5 Fractal Display

Plate 12. Cosine Fractal

Plate 13. The Third-Order Newton's Method Fractal

Plate 14. A Plasma Mountain

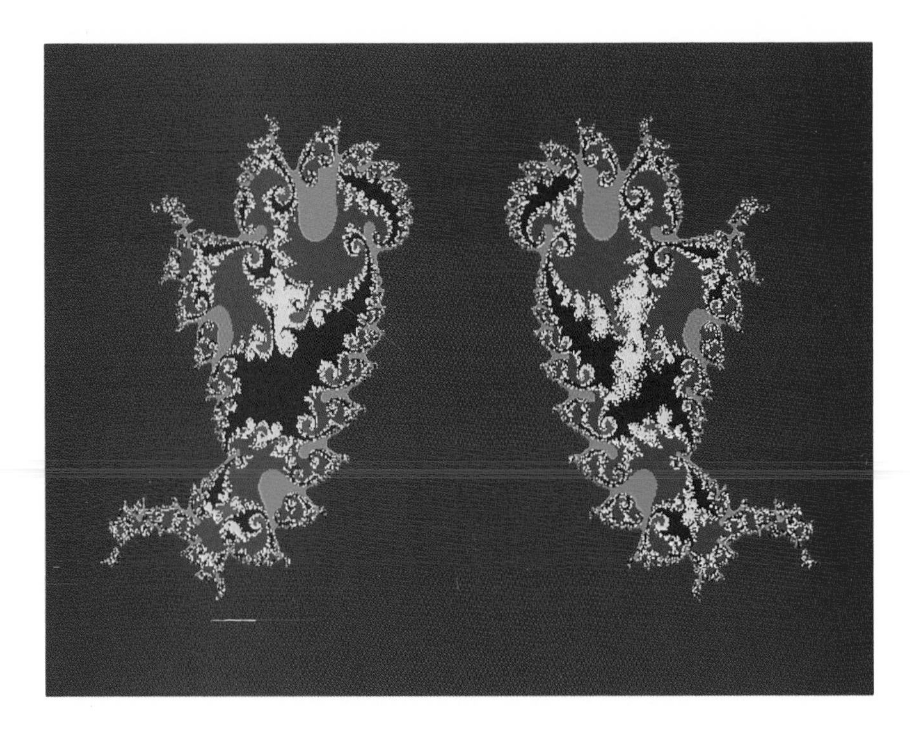

Plate 15. The Phoenix Fractal

Plate 16. The Seventh-Order Newton's Method Fractal

```
        color = random(dist<<1) - dist;
        color += (readPixel(x1,y1) + readPixel(x2,y1) + 1)
              >> 1;
        color = (color < 1) ? 1: (color > 255) ? 255: color;
        if (readPixel(x,y1) == 0)
              plot(x,y1,color);
        color = random(dist<<1) - dist;
        color += (readPixel(x1,y2) + readPixel(x2,y2) + 1)
              >> 1;
        color = (color < 1) ? 1: (color > 255) ? 255: color;
        if (readPixel(x,y2) == 0)
              plot(x,y2,color);
        dist = y2 - y1;
        color = random(dist<<1) - dist;
        color += (readPixel(x2,y1) + readPixel(x2,y2) + 1)
              >> 1;
        color = (color < 1) ? 1: (color > 255) ? 255: color;
        if (readPixel(x2,y) == 0)
              plot(x2,y,color);
        color = random(dist<<1) - dist;
        color += (readPixel(x1,y1) + readPixel(x1,y2) + 1)
              >> 1;
        color = (color < 1) ? 1: (color > 255) ? 255: color;
        if (readPixel(x1,y) == 0)
              plot(x1,y,color);
        color = (readPixel(x1,y1) + readPixel(x2,y1) +
              readPixel(x2,y2) + readPixel(x1,y2) + 2) >> 2;
        plot(x,y,color);
        subdivide(x1,y1,x,y);
        subdivide(x,y1,x2,y);
        subdivide(x,y,x2,y2);
        subdivide(x1,y,x,y2);
}
```

13

Structures and Unions

Structures provide a vehicle for defining a single variable that actually consists of a combination of several variables. These component variables do not even have to be of the same type. The entire structure may be passed to a function or returned from a function as a single entity. This gets around one of the principal limitations of a function. Unions enable the same area of memory to be defined for several different data types. This could be used, for example, to send a number of integers to an array and then transfer this data to a disk on a byte-by-byte basis.

Structures

The following is a typical structure definition:

```
struct struct_id
{
        int x;
        int y;
        float a;
        char b;
};
```

This is the only situation in C where a pair of curly brackets *{}* is followed by a semicolon. The structure is a type, so that you could write

```
struct struct_id
{
        int x;
        int y;
        float a;
        char b;
} c, d;
```

and you would end up with two arguments, *c* and *d* each of which is of the type defined by the preceding structure. The structure nomenclature, given in the examples by the term *struct_id* is optional. If you put all of your arguments that are of this structure type at the end of the *struct* statement, as are *c* and *d* in the second example, you don't need any structure nomenclature term at all; if you don't do this, you need to name the structure so that you can use the name to identify terms having this structure later with a statement such as

```
struct struct_id f;
```

This will give you a new argument *f* of the type defined by the *struct* statement of which *struct_id* is the nomenclature.

The *typedef* Statement

The *typedef* statement enables you to define your own data types. For example

```
typedef char CHARACTER;
```

defines a data type *CHARACTER*, which is the same as the predefined data type *char*. We can also use *typedef* with a structure. We would have a statement like this

```
typedef struct
{
      int x;
      int y;
      float a;
      char b;
} struct_id;

struct_id f;
```

The name *struct_id* now has a different meaning. It is the name of a data type rather than the name of a structure. When *struct_id* is the name of a structure type, to define an argument that has this structure, we must first use the word *struct*, then follow it by the name given to the structure type (*struct_id* in the example), followed by the variable that is to be of this type (*f*), so that we have

```
struct struct_id f;
```

When *struct_id* is the name of the data type we define an argument of this data type in just the same way that we define the type of any argument, namely using the name of the data type (*struct_id*) followed by the argument name (*f*). Thus we have

```
struct_id f;
```

Regardless of which method we use to define the argument (*f*), we use *f* in just the same way throughout the program.

Using a Structure

Now that we have a structure, such as one of the examples given previously, how do we use it? We are concerned with two different things. First, we are interested in handling individual arguments that make up the structure. Thus we might have

```
f.x = 2;
f.y = 3;
f.a = 3.14159;
f.b = 'T';
```

For this example, *f* is an argument having the structure given by *struct_id* in the previous examples. You'll remember that this structure consisted of integers *x* and *y*, floating point number *a* and character *b*. The lines we just gave refer to these components of argument *f*. The expressions *f.x, f.y, f.a*, and *f.b* refer to the *x, y, a*, and *b* components of *f*, respectively. The period operator is preceded by the argument name and is followed by the name of the variable in the structure with which we want to work. The lines of code just given assign a value to each member of the structured argument *f*.

The other way in which we wish to treat a structure is as a single entity. For example:

```
struct_id function_name(struct_id s);
```

is the prototype of a function that both passes and returns an argument of type *struct_id*. By passing the one argument, we have made all of the components of the structure available to the function. By stating that the function is of the type *struct_id*, we return all of the components that make up the structure in one operation. We can thus do the following:

```
struct_id f,g;
.
.
g = function_name(f);
.
.

struct_id function_name(struct_id s)
{
      struct_id v;

      v.x = 2 * s.x;
      v.y = 2 * s.y;
      v.a = 2.0 * s.a;
      v.b = s.b;
      return(v);
}
```

The argument *g* in the main program now has its components all set up for the values that the function inserted into the structure *v*.

Unions

The *union* statement defines a section of memory in which different types of data overlap. The syntax is very similar to that of a structure. For example,

```
union union_id
{
      char a;
      int b;
} D;
```

What happens with the union is much different than with a structure, however. First, every argument in the union begins at the same memory address, so they all overlap. No matter how many arguments you have in a union, they all overlap each other. The amount of memory allocated to the union is that of the largest argument of which it is made up. Thus, in the example, two bytes of memory would be allocated to hold an integer. If you specified an operation on *D.b*, these two bytes would contain the integer value that you are using. If you specified an operation on *D.a*, the character that you are operating on would be the least significant byte of *D.a* since all arguments that make up the union overlap in memory. Remember that every argument in the union overlaps every other argument.

Therefore, if you want to have two **sets** of data overlap, you must make each set a structure and then have a union of the two structures. In Chapter 4, you encountered the union *REGS* which stores data for the microprocessor registers, either when they are accessed as one-byte or two-byte registers. Let's look at how this is done in the C compiler:

```
struct WORDREGS
{
      unsigned int    ax, bx, cx, dx, si, di, cflag,
            flags;
};

struct BYTEREGS
{
      unsigned char   al, ah, bl, bh, cl, ch, dl, dh;
};

union   REGS
{
      struct  WORDREGS x;
      struct  BYTEREGS h;
};
```

Let's look at this in a little further detail. Since the union has the two structures *WORDREGS* and *BYTEREGS*, the larger (which is *WORDREGS*) determines the size of the memory location, which is eight integers, or 16 bytes. The structure defines eight integer length registers, whose contents may then be passed to the microprocessor registers when the union is used. We often want to access individually either the most significant or least significant byte of each of the first four of these registers. For example, for one application we might want to access just the most significant byte (*ah*) of the *ax* register, whereas for another application we might want to access the least significant byte (*al*) of the *ax* integer. This is accomplished through the structure *BYTEREGS*, which overlaps the first 8 bytes of the union and assigns the proper register names. (Note that the 8086 family of microprocessors orders each word with the least significant byte first and then the most significant byte, which is why the register order is *al, ah*, rather than the other way around.) Now suppose we have defined the argument

```
union REGS reg;
```

We can then assign a value with

```
reg.x.ax = 0x4532;
```

or we can accomplish the same result by

```
reg.h.ah = 0x45;
reg.h.al = 0x32;
```

This is a good example of how a union can be used effectively. Another use occurs when we want to read a number of different types of data from a disk file. We could read each piece of data separately, but this requires a lot more time than if we were to access the disk file only once. Therefore, we set up a union of which one component is a buffer array having the character length of all the data we want to read from the file in one operation. The other component of the union is a structure made up of all the data types we want to read in the proper sequence. We then read the proper number of bytes into the buffer defined by the first union member in a single operation. Then, by using the second union component, we have access to each individual data item.

Expanding the Tchebychev C_5 Fractal

In Chapter 10, we described the Tchebychev C_5 fractal and showed how to save and restore the display using a disk file. In this chapter, we're going to look at this fractal again, giving a program that will start with the display of the entire Tchebychev C_5 fractal that was produced by the program of Chapter 10 and allow us to select an expansion of it and produce the expanded display. This program will be written using a lot of structures, so as to give you some familiarity with how these are used within an actual program. The program to expand the Tchebychev C_5 fractal is shown in Figure 13-1. After expanding twice, we have the picture shown in Plate 11. The coordinates of the top left-hand corner of this expanded display are $z.x = -0.210435$ and $z.y = 0.050792$ and the coordinates of the lower right-hand corner are $z.x = -0.205171$ and $z.y = 0.044194$.

At the beginning of the program, we define two *typedef*s, which are structures. The first, *complex*, consists of two floating point numbers where x is the real component of a complex number and y is the imaginary component. The second structure is *boundary*, which consists of two complex numbers. The first, *max*, represents the coordinates of the top left-hand corner of a display and the second, *min*, represents the coordinates of the bottom right-hand corner of the display.

Now look at the main program. We start by evaluating *five*, which is just the *complex* representation of the number 5. Next we set the value for each component of *bounds*, which is of the type *boundary*. These bounds are those that were used for the Tchebychev polynomial C_5 which you created and saved in Chapter 10. (If you haven't done this, stop and do it now, since we need the saved file for the Tchebychev display to use with this program.) Next, *readscrn* is used to display the Tchebychev fractal. (The function *readscrn* was described in Chapter 10.) The program then enters a *while* loop, which repeats as long as *ch1* is not the *Ent* key. (The argument *ch1* was initialized to be 0, so we make at least one pass through this loop.) The program next computes the increment by which the complex number values are changed for each pixel shift in the row or column values. We begin by using the *sub* function to subtract the minimum from the maximum values of the boundaries to get the differences. (Near the end of the listing you will find the *add*, *sub*, and *mul* functions, which add, subtract, and multiply, respectively, two complex numbers.) The differences are divided by the number of pixels in a row and column, respectively. Now we set *bounds* equal to the return from the function *move_cursor*. This function, which will be described next, allows you to select the corners of a new, expanded display and returns the boundaries representing these corners. The program then clears the screen, preinitializes the complex number imaginary value for each row, and then enters three nested *for* loops that generate the fractal picture. The outer loop iterates once for each column. At the start of each iteration, it sets the value for the real part or the complex number *c*. The next loop iterates once for each row. At the start of each iteration, it sets the initial values of *z.x, z.y, c.y, oldz.x*, and *oldz.y*. Together, these two loops cover every pixel on the screen. Next comes the innermost *for* loop, which actually iterates the fractal equation for each pixel. You should compare the mathematics within this loop with that given for the separate components in Chapter 10 and note the differences that result from using structures. This loop also uses the check for recycling that was used with the Mandelbrot set in Chapter 7 to speed up operation. After all of the loops are complete, the program pauses until a key is struck. When this occurs, the key information is read into *ch1*. If the key struck was the *Ent* key, the program breaks out of the overall *while* loop and terminates. If any other key is struck, the whole process is repeated again, with an opportunity to create a further expanded display that is a selected portion of the expanded display that just was created.

The *move_cursor* Function

This function makes use of the cursor arrow keys to move first an upper left corner and then a lower right corner to mark out the new rectangle that is to form the expanded display. The cursor arrow keys, when struck, return two bytes to the keyboard buffer. The first byte for each of these keys is a NULL (00). When a NULL is read, the *getch* function is used to read the second byte and 256 is added to it to give a unique number. The *move_cursor* function begins by defining the arrow keys by the numbers that will be created by this process. When the *move_cursor* function starts, the parameter *type* is initialized to zero. The function begins with a *while* loop that iterates as long as *type* is less than 2. On the first pass, an *if* statement causes the function to draw an upper left corner. The function then enters an inner *while* loop that iterates as long as the key that is read is not the *Ent* key. Once a key is struck, the corner is rewritten, which since it is written each time in the exclusive-OR mode, causes the previously displayed corner to be erased. The function then enters a *switch* statement, which checks for one of the arrow keys. The cursor position is moved one pixel in the arrow direction, provided the boundaries of the display are not exceeded. The cursor corner is redisplayed at the new position. The coordinates of the corner are displayed and placed in the appropriate members of the argument *limits*, which is of the *boundary* type. The upper left corner cursor continues to be moved by the cursor arrow keys until the *Ent* key is struck. At this point, the *type* parameter is incremented and the cursor coordinates are each increased by 16. The next pass through the outer *while* works in the same way. The last upper left corner cursor that was displayed is left on the screen, and for the second pass, a new lower right corner cursor is displayed. This is moved with the cursor arrow keys and its values displayed and stored in the *limits* structure. When the *Ent* key is hit again, the *move_cursor* function ends, passing the values for the new display boundaries back to the calling program.

The *plot* Function

The plot function used here is a little different than we have encountered before. In addition to the position and colors, it is passed a *type* parameter. If this parameter is 0, the function just plots a point in the selected color at the designated location. This is done using a combination of outputs to the graphics controller registers and the sequence registers rather than to the graphics controller registers alone, as has been done with previous *plot* functions. This value of *type* is used for plotting the points on the fractal display. When *type* is 1, an additional register output occurs that causes the pixel to be written in the exclusive-OR mode. This is used to display the corner cursors, since by writing over the same pixel with another exclusive-OR, the original display data is returned, effectively erasing the cursor and restoring the original display.

Figure 13-1. Listing of Program to Expand Tchebychev C_5 Fractal

```
/*

         TCHEBEX = Program to generate expanded Tchebychev
                        C5 fractal set.

                    By Roger T. Stevens  7-13-92

*/

#include <dos.h>
#include <stdio.h>
#include <math.h>

typedef struct
{
      float x;
      float y;
} complex;

typedef struct
{
      complex max;
      complex min;
} boundary;

boundary bounds;
complex c, z, z2, z3, z4, z5, delta, MAX, MIN, five, oldz;
```

(continued)

```
int xres = 640, yres = 480;
int  color, row, col, ch1 = 0;
float Q[480];
complex add(complex a, complex b);
complex sub(complex a, complex b);
boundary move_cursor(void);
complex mul(complex a, complex b);
void cls(int color);
void plot(int x,int y,int color, int type);
void readscrn(void);
void setmode(int mode);

int i;

main()
{
      five.x = 5.0;
      five.y = 0.0;
      bounds.max.x = 1.0;
      bounds.min.x = -1.0;
      bounds.max.y = 0.4;
      bounds.min.y = -0.4;
      readscrn();
      while (ch1 != 0x0D)
      {
            delta = sub(bounds.max, bounds.min);
            delta.x /= xres;
            delta.y /= yres;
            bounds = move_cursor();
            delta = sub(bounds.max, bounds.min);
            delta.x /= xres;
            delta.y /= yres;
            cls(7);
            for (row=0; row<yres; row++)
                  Q[row] = bounds.max.y - row*delta.y;
            for (col=0; col<xres; col++)
            {
                  c.x = bounds.min.x + col*delta.x;
                  for (row=0; row<yres; row++)
                  {
                        z.x = 0.5;
                        z.y = 0.0;
                        c.y = Q[row];
                        oldz.x = 0;
                        oldz.y = 0;
                        for (color=0; ((color<64) &&
                              ((z.x*z.x + z.y*z.y) <
                              1000)); color++)
                        {
                              z2 = mul(z,z);
                              z3 = mul(z2,z);
                              z4 = mul(z2,z2);
                              z5 = mul(z4,z);
```

(continued)

```
                              z3 = mul(z3,five);
                              z = mul(z,five);
                              z = add(z5,z);
                              z = sub(z,z3);
                              z = mul(c,z);
                              if ((z.x == oldz.x) && (z.y
                                   == oldz.y))
                         {
                              color = 0;
                              break;
                         }
                         if ((color % 8) == 0)
                         {
                              oldz.x = z.x;
                              oldz.y = z.y;
                         }

                         }
                         plot(col, row, color%16,0);
                    }
               }
               ch1 = getch();
          }
}

/*
     plot() = Plots a point on the screen at a designated
          position using a selected color for 16 color
             modes using either overwriting or XOR.
*/

void plot(int x, int y, int color, int type)
{
     #define graph_out(index,val)  {outp(0x3CE,index);\
                              outp(0x3CF,val);}
     #define seq_out(index,val)  {outp(0x3C4,index);\
                              outp(0x3C5,val);}

     int dummy,mask;
     char far * address;

     address = (char far *) 0xA0000000L + (long)y *
          xres/8L + ((long)x / 8L);
     mask = 0x80 >> (x % 8);
     graph_out(8,mask);
     if (type == 1)
          graph_out(3,24);
     seq_out(2,0xFF);
     dummy = *address;
     *address = 0;
```

(continued)

```
        seq_out(2,color);
        *address = 0xFF;
        seq_out(2,0x0F);
        graph_out(3,0);
        graph_out(8,0xFF);
}

/*
        move_cursor() = Moves the cursor on the screen.
*/

boundary move_cursor(void)
{
        #define LEFT_ARROW              331
        #define RIGHT_ARROW             333
        #define UP_ARROW                328
        #define DOWN_ARROW              336
        int i, ch=0, x=320, y=240, type=0, min_x=16,
              min_y=16;
        complex temp;
        boundary limits;

        while (type < 2)
        {
              if (type == 0)
              {
                    for (i=0; i<16; i++)
                          plot(x+i,y,15,1);
                    for (i=0; i<16; i++)
                          plot(x,y+i,15,1);
              }
              else
              {
                    for (i=-15; i<=0; i++)
                          plot(x+i,y,15,1);
                    for (i=-15; i<=0; i++)
                          plot(x,y+i,15,1);
              }
              while (ch != 0x0D)
              {
                    ch = getch();
                    if (ch == 0x00)
                          ch = getch() + 256;
                    if (type == 0)
                    {
                          for (i=0; i<16; i++)
                                plot(x+i,y,15,1);
                          for (i=0; i<16; i++)
                                plot(x,y+i,15,1);
                    }
```

(continued)

```
else
{
      for (i=-15; i<=0; i++)
            plot(x+i,y,15,1);
      for (i=-15; i<=0; i++)
            plot(x,y+i,15,1);
}
switch(ch)
{
      case UP_ARROW:
            if (y>min_y)
                  y--;
            break;
      case DOWN_ARROW:
            if (y<472)
                  y++;
            break;
      case LEFT_ARROW:
            if (x>min_x)
                  x--;
            break;
      case RIGHT_ARROW:
            if (x<632)
                  x++;
            break;
}
if (type == 0)
{
      for (i=0; i<16; i++)
            plot(x+i,y,15,1);
      for (i=0; i<16; i++)
            plot(x,y+i,15,1);
}
else
{
      for (i=-15; i<=0; i++)
            plot(x+i,y,15,1);
      for (i=-15; i<=0; i++)
            plot(x,y+i,15,1);
}
gotoxy(1,1);
if (type==0)
{
      limits.max.y = bounds.max.y -
            y*delta.y;
      limits.min.x = bounds.min.x +
            x*delta.x;
      printf("z = (%f, + i %f)"
            ,limits.min.x,
            limits.max.y);
}
else
{
```

(continued)

```
                              limits.min.y = bounds.max.y -
                                    y*delta.y;
                              limits.max.x = bounds.min.x +
                                    x*delta.x;
                              printf("z = (%f, + i %f)"
                                    ,limits.max.x,
                                    limits.min.y);
                        }
                  }
                  ch = 0;
                  type++;
                  x+=16;
                  y+=16;
            }
            return(limits);
}

/*
      readscrn() = Reads and displays a graphics disk file.
*/

void readscrn(void)
{
      struct SREGS segregs;
      FILE *fsave;
      char buffer[38400];
      int i, srcseg, srcoff;

      setmode(18);
      cls(7);
      segread(&segregs);
      srcseg = segregs.ds;
      srcoff = (int) buffer;
      fsave = fopen("tcheb.raw","rb");
      for (i=0; i<4; i++)
      {
            fread(buffer,1,38400,fsave);
            outport(0x3C4,(0x01<<(8+i)) + 2);
            movedata(srcseg,srcoff,0xA000,0,38400);
      }
      fclose(fsave);
}

/*
                  setmode() = Sets video mode
*/
```

(continued)

```
void setmode(int mode)
{
      union REGS reg;

      reg.x.ax = mode;
      int86 (0x10,&reg,&reg);
}

/*
                  cls() = Clears the screen
*/

void cls(int color)
{
       union REGS reg;

       reg.x.ax = 0x0600;
       reg.x.cx = 0;
       reg.x.dx = 0x1E4F;
       reg.h.bh = color;
       int86(0x10,&reg,&reg);
}

/*
      plot() = Plots a point on the screen at a designated
          position using a selected color for 16 color
                            modes.
*/

void plot(int x, int y, int color, int type)
{
      #define graph_out(index,val)  {outp(0x3CE,index);\
                                 outp(0x3CF,val);}
      #define seq_out(index,val)  {outp(0x3C4,index);\
                                 outp(0x3C5,val);}

      int dummy,mask;
      char far * address;

      address = (char far *) 0xA0000000L + (long)y *
            xres/8L + ((long)x / 8L);
      mask = 0x80 >> (x % 8);
      graph_out(8,mask);
      if (type == 1)
            graph_out(3,24);
      seq_out(2,0xFF);
```

(continued)

```
        dummy = *address;
        *address = 0;
        seq_out(2,color);
        *address = 0xFF;
        seq_out(2,0x0F);
        graph_out(3,0);
        graph_out(8,0xFF);
}

/*
                    add() = Adds two complex numbers
*/

complex add(complex a, complex b)
{
        complex c;

        c.x = a.x + b.x;
        c.y = a.y + b.y;
        return c;
}

/*
                    sub() = Subtracts two complex numbers
*/

complex sub(complex a, complex b)
{
        complex c;

        c.x = a.x - b.x;
        c.y = a.y - b.y;
        return c;
}

/*
                    mul() = Multiplies two complex numbers
*/

complex mul(complex a, complex b)
{
        complex c;
```

(continued)

```
        c.x = a.x*b.x - a.y*b.y;
        c.y = a.x*b.y + a.y*b.x;
        return c;
}
```

14

Mathematical Functions

You're not likely to have much trouble using the mathematical functions that are included in most C compiler libraries. They are inserted into your code in the same way that they would be included in mathematical equations. For example,

```
a = cos(b);
```

will set the variable *a* to the cosine of the angle *b*. (Note, however, that *b* is the angle in radians, not in degrees.) There are just a few things to note to avoid any mistakes.

1. Be sure that the proper *#include* statement is at the beginning of your program. For most mathematical functions, this is

```
#include <math.h>
```

However, some compilers have a lot of new mathematical functions which may be in different libraries. Be sure to check. The *#include* statement references the header file that lists the functions you want to use in your program and permits the compiler to link up the appropriate library functions. The least horrible thing that will happen if you don't have the right header in an *#include* statement is that during compilation the compiler will give you an error message stating that the function you are calling does not exist. The most horrible thing (which often occurs when the *math.h* header is not referenced) is that everything will proceed normally, but when you run your program, all cosines, for example, will be set to 0, regardless of the value they should have. The results are always disastrous, and it often takes quite awhile to realize what is happening.

2. Be sure that you limit your inputs to functions to the acceptable range of values, if a range of values is specified. For example,

```
theta = asin(anglesin);
```

says that *theta* is the angle whose sine is *anglesin*. The allowable limits for *anglesin* are -1.0 to +1.0 (since mathematically the sine must be within these limits). It is your responsibility as programmer to make sure that *anglesin* can never take on values outside these limits. If you don't do this, when the program encounters a value outside the limits, an error message will be produced and your whole program will grind to a halt.

3. All angles used in C mathematical functions are in radians. If you are uncomfortable with this and prefer degrees, you can easily create your own functions. Here is an example of how simple it is to make such a function. This one gives the cosine of an angle expressed in degrees.

```
double COS(double angle)
{
      double temp;

      temp = cos(0.017453292*angle);
      return temp;
}
```

Table 14-I lists the common mathematical functions that are supplied with C compilers, together with their descriptions and any limitations on their inputs and outputs.

The Cosine Fractal

The cosine fractal program makes use of the *cos* and *cosh* functions and therefore will give you some idea of how simple it is to use mathematical functions in actual programming. Each pixel on the screen represents the location of a particular complex number *c* where the left edge of the screen is the minimum value of the real part of *c* *(Pmin)*, the right-hand edge of the screen is the maximum value of the

Table 14-I. Mathematical Functions in C

Function	Description	Input Limits	Output Limits
abs	Returns the absolute value of an integer.	-32,767 to 32,767	0 to 32,767
acos	Returns the angle (in radians) whose cosine is input.	-1.0 to 1.0	0 to pi
asin	Returns the angle (in radians) whose sine is input.	-1.0 to 1.0	-pi/2 to pi/2
atan	Returns the angle (in radians) whose tangent is input.	none	-pi/2 to pi/2
atan2	Returns the angle (in radians) whose tangent is y/x. Format: `c = atan2(y,x);`	x and y cannot both be 0	-pi to pi
ceil	Returns the smallest integer not less than the floating point number input.	none	none
cos	Returns the cosine of the angle (in radians) input.	none	-1.0 to 1.0
cosh	Returns the hyperbolic cosine $(e_x + e^{-x})/2$, of the number input.	none	none
div	Returns the quotient and remainder where numerator and denominator are input. `x = (numer,denom);` where *x* must be of the type *div_t*.	none	none

(continued)

Table 14-I. Mathematical Functions in C (cont.)

Function	Description	Input Limits	Output Limits
exp	Returns value of exponential *e* raised to the power input.	none	none
fabs	Returns the absolute value of a floating point number.	none	none
floor	Returns the largest integer not greater than the floating point number input.	none	none
fmod	Returns *x* modula *y*, the remainder of *x*/*y* for floating point numbers *x* and *y*.	none	none
frexp	Returns mantissa of input and stores exponent of input in *exponent*. `double frexp(double input,int *exponent)` where $input = m \times 2^n$.	none	none
hypot	Returns the smallest integer not less than the floating point number input.	none	none
log	Returns the natural logarithm of the input. (Note that the natural logarithm is usually written *ln* in mathematical equations.)	input cannot be 0.	none

(continued)

Table 14-I. Mathematical Functions in C (cont.)

Function	Description	Input Limits	Output Limits
log10	Returns the logarithm to the base 10 of the input. (Note that the logarithm to the base 10 is usually written *log* in mathematical equations.)	input cannot be 0	none
modf	Returns the fractional part of a *double* input. The integer part is stored in *ipart*. `double modf(double x, *double ipart).`	none	none
pow	Returns the result of raising the first input to the second input power. `double pow(double x, double y).`	if x is less than 0, then y must be a positive integer	none
pow10	Returns 10 to the power of the input.	none	none
rand	Returns a pseudorandom number.	---	0 to 32,767
sin	Returns the sine of the angle (in radians) input.	none	-1.0 to 1.0
sinh	Returns the hyperbolic sine, $(e^x - e^{-x})/2$, of the number input.	none	none

(continued)

Table 14-I. Mathematical Functions in C (cont.)

Function	Description	Input Limits	Output Limits
sqrt	Returns the positive square root of the input.	real, non-negative	none
srand	Reinitializes the random number generator with the seed number input. Each run having a given seed will produce the same series of pseudorandom numbers.	0 to 65,535	---
tan	Returns the tangent of the angle (in radians) input.	none	none
tanh	Returns the hyperbolic tangent *(sinh x)/(cosh x)* of the number input.	none	none

real part of *c (Pmax)*, the bottom edge of the screen is the minimum value of the imaginary part of *c (Qmin),* and the top edge of the screen is the maximum value of the imaginary part of *c (Qmax)*. For each value of *c* we iterate the equation

$$z_{n+1} = \cos(z_n) + c$$

(Equation 14-1)

where *z* is also a complex number (starting with $z_0=0$) until the sum of the squares of the real and imaginary parts of *z* is greater than or equal to 100. At this point, we stop and color the pixel represented by the current value of *c* in accordance with the number of iterations that have taken place. (You may remember from your trig days that the cosine is limited to values between -1 and +1 and wonder how the condition just described can ever occur. However, these limits to the cosine apply only to the cosine of real angles, which are what we normally work with. In this equation, we are taking the cosine of complex angles, so the limits don't apply.) You may be lucky enough to have a compiler such as Borland C++ whose library includes trigonometric functions of complex numbers. If so, you can simplify the math in the program a lot. The program as written takes care of compilers that can't handle complex numbers by using the trigonometric identity

$$\cos(x + iy) = \cos x \cosh y - i \sin x \sinh y \qquad \text{(Equation 14-2)}$$

This permits you to represent the complex functions using ordinary trigonometric functions available with any C compiler.

The program is listed in Figure 14-1. The first thing to note is that in order to achieve the most efficiency, we need to precompute all possible values of Q and store them in an array. The main part of the program consists of three nested *for* loops. The outermost loop iterates once for each column of the display screen. This loop begins with the column designator *col* set to 0. When *col* reaches *xres* we have reached the right-hand edge of the screen and the loop terminates. The loop begins by setting up the proper value of P, the real part of the complex number c in Equation 14-1. The next *for* loop iterates once for each row on the display screen. It begins with the row designator *row* set to 0. When *row* reaches *yres*, we have reached the bottom of the screen and the loop terminates. Together, these two loops cover every pixel location that exists on the display screen.

The innermost *for* loop does all the work of solving the iterated equation. Just before entering the loop, the parameters x and y are initialized to 0. The test condition for continuing in the loop is a little complicated. For another iteration of the loop to occur, two things must be true. First, the iteration number (value of *color*) must be less than 64. (The parameter *color* is incremented on each pass through the loop.) Second, the value of the sum of the squares of x and y (the real and imaginary parts of z) must be less than 1000. When either of these conditions becomes false, the loop terminates. Mathematically, when the second condition becomes false, we know that continued iteration of the loop is going to cause the value of z to blow up (go to infinity). If the first condition becomes false, we know we have completed the number of iterations specified with z still remaining well-behaved. It is either growing very very slowly or has reached some sort of stable condition. This might either be that it settles on some fixed value or that it is cycling through a few values over and over. In either case, we don't want to continue iterating any further. Within this loop is code that checks for these stable conditions and, if it finds one of them, exits the loop without continuing to the maximum number of iterations specified.

The mathematics within the inner loop uses the *cos* and *cosh* functions together with some other mathematical operations to solve Equation 14-1 for each pixel location. The point is then plotted to the screen in the color determined by the number of iterations of the equation that occurred. The program generates the very lovely fractal cosine pattern shown in Plate 12. You can change the values of *Pmax, Pmin, Qmax*, and *Qmin* to expand or contract this pattern or to select another part of the cosine fractal for display. If you don't have a math coprocessor in your computer, the program will run quite slowly, but the result is worth waiting for and watching the actual creation of the display is quite interesting.

Figure 14-1. Listing of Program to Generate Cosine Fractal

```
/*

    COSFRAC = Program to generate cosine fractal set

              By Roger T. Stevens  1-22-92

*/

#include <dos.h>
#include <stdio.h>
#include <math.h>

int xres = 640, yres = 480;
int  color, row, col;
double Pmax=1.05, Pmin= 0.92, Qmax = 1.8, Qmin = 1.6;
double Q[480], P, deltaP, deltaQ, old_x, old_y, temp, x,
      y, xsq, ysq;

void plot(int x,int y,int color);
void setmode(int mode);
void cls(int color);

main()
{
      setmode(18);
      cls(7);
```

(continued)

```
        deltaP = (Pmax - Pmin)/xres;
        deltaQ = (Qmax - Qmin)/yres;
        row = -1;
        while (++row<yres)
                Q[row] = Qmax - row*deltaQ;
        for (col=0; col<xres; col++)
        {
                P = Pmin + col*deltaP;
                for (row=0; row<yres; row++)
                {
                        x = y = 0.0;
                        for (color=0; ((color<64) && ((x*x +
                                y*y) < 1000)); color++)
                        {
                                temp = cos(x)*cosh(y) + P;
                                y = -sin(x)*sinh(y) + Q[row];
                                x = temp;
                        }
                        if (color >= 64)
                                color = 0;
                        else
                                color = color % 15 + 1;
                        plot(col, row, color);
                }
        }
        getch();
}

/*
┌──────────────────────────────────────────────────┐
│                                                  │
│          setmode() = Sets video mode             │
│                                                  │
└──────────────────────────────────────────────────┘
*/

void setmode(int mode)
{
        union REGS reg;
        reg.x.ax = mode;
        int86 (0x10,&reg,&reg);
}

/*
┌──────────────────────────────────────────────────┐
│                                                  │
│          cls() Clears the screen                 │
│                                                  │
└──────────────────────────────────────────────────┘
*/
void cls(int color)
{
        union REGS reg;
```

(continued)

```
        reg.x.ax = 0x0600;
        reg.x.cx = 0;
        reg.x.dx = 0x1E4F;
        reg.h.bh = color;
        int86(0x10,&reg,&reg);
}

/*

    plot() = Plots a point on the screen at a designated
        position using a selected color for 16 color
                        modes.

*/

void plot(int x, int y, int color)
{
      #define graph_out(index,val)  {outp(0x3CE,index);\
                                 outp(0x3CF,val);}

      int dummy,mask;
      char far * address;

      address = (char far *) 0xA0000000L + (long)y *
            xres/8L + ((long)x / 8L);
      mask = 0x80 >> (x % 8);
      graph_out(8,mask);
      graph_out(5,2);
      dummy = *address;
      *address = color;
      graph_out(5,0);
      graph_out(8,0xFF);
}
```

15

Character and Number Conversions

As you begin to develop more complex C programs, you will often encounter the need to convert a string of ASCII characters to an integer or floating point number, or conversely to convert one of these numbers to a string of ASCII characters. If you think a little bit about the code that you would have to write to perform such a conversion, you will be forever thankful that functions to do these conversions are a part of the library included with every C compiler.

The *atoi* Function

The *atoi* function converts a string of ASCII numbers (characters in the range 0x30 to 0x39) to an integer. The function is passed a pointer to a string and returns an integer. If a non-numeric character is encountered in the string, the conversion process ends. The function will ignore tabs or spaces preceding the numerical part of the string, but a tab or space imbedded in the number will stop the conversion process. The function will also recognize a minus sign preceding the first number. Here is a simple demonstration program

```
#include <stdio.h>
#include <stdlib.h>
#include <conio.h>

void main(void)
{
      int n;
```

(continued)

```
        char string1[8] = {"5432"}, string2[8] =
              {"123K456"};

        printf("\nstring 1 = %s",string1);
        n = atoi(string1);
        printf("\ninteger = %d", n);
        printf("\nstring 2 = %s",string2);
        n = atoi(string2);
        printf("\ninteger = %d", n);
        getch();
}
```

For the first string, *5432*, the integer output is *5432*. For the second string, *123K456*, the integer output is *123* since the conversion ends when the improper character *K* is encountered. You also need to keep the value represented by the string within the acceptable limits for an integer (-32768 to +32767). If the number is greater than +32767 and less than 65536, you will find it converted to a complimentary negative number, is probably not what you want. Similarly numbers between -32768 and -65537 will convert to complimentary positive numbers. For numbers even greater or smaller than the limits, the number will be the string value modula 65536 converted as given above. In any case, numbers outside the limits are to be avoided. It's your job as programmer to assure that such cases do not occur.

The *atol* Function

The *atol* function is just the same as the *atoi* function except that it returns a long integer rather than an ordinary integer. You can still get in trouble by having a string that exceeds the allowable limits for a long integer, but your chances for trouble are much less likely since the limits for a long integer are so much wider than for an ordinary integer.

The *atof* and *_atold* Functions

The *atof* function accepts as an argument the address of a string of ASCII characters representing a floating point number and returns the corresponding number as type *double*. The form of the string is

[spaces or tabs][sign][digits][.][digits][e or E][sign][digits]

Each of these individual elements is optional, but of course you must have something in the string for the function to convert. As the first element, you may have as many spaces or tabs as desired at the beginning of the string. Next, you may have a sign; if none is present, the number will be assumed to be positive. Next come any digits (ASCII characters between 0x30 and 0x39) that are to precede the decimal point. Next comes the decimal point; if it is missing, it will be assumed to be at the end of the string of digits. Next, optionally, comes the letter *e* or *E* to indicate the exponential form, meaning that the sign and digits following this letter represent a power of 10 by which the number is to be multiplied. [The *e* or *E* must directly follow the preceding digits without any space and it must be followed without intervening spaces by the sign (if used) and digits that make up the power of 10.] Next comes an optional sign. If it is negative, the number is divided by the following power of 10; if it is positive or is not used, the number is multiplied by the following power of 10. Finally come the digits that represent the power of 10 to be used. Here are a few legitimate strings that can be converted by *atof*:

```
"3.14159625"
"     123.456"
"-65.789"
"123.456E-23"
"   -26.224e5"
"   +567.896E+6"
"+NAN"
"+INF"
```

The last two strings may be either plus or minus. They are interpreted as meaning *not a number* and *infinity*, respectively, and are passed to the floating point number.

The *_atold* function is just the same as the *atof* function except that it returns a number of type *long double* rather than *double*. Before attempting to use this function, make sure that *long double* types are supported by your compiler.

The *strtod* and *_strtold* Functions

The *strtod* and *_strtold* functions are just like the *atof* and *_atold* functions, respectively, as far as their conversion actions are concerned. They do, however, have an additional error detecting capability that is useful if an unacceptable character is encountered in the string. The method used is a little obscure. Take a look at the following code fragment:

```
char *endptr;
char string[34];
double number;
...
...
number = strtod(string,&endptr);
```

What happens here is that the second argument passed to the *strtod* function is a pointer to a pointer. When the program returns from the *strtod* function, the pointer *endptr* is a NULL if the conversion process was completed successfully. If an unacceptable character was encountered in the string, then *endptr* contains something other than a NULL. What it contains is, in fact, a pointer to a memory location that contains the address of the character in the string that stopped the conversion process. This can be useful in detecting unacceptable character strings. We'll show how it's used in the section below that deals with Newton's method fractals.

The *itoa* Function

The *itoa* function is the exact opposite of the *atoi* function. It accepts an integer argument and converts it into a null-terminated string of ASCII characters. A typical use of the function would look like this

```
itoa(number, string, 10);
```

The argument *number* is the integer that is input to the function. The second argument, *string*, is the address of the character array that will contain the ASCII string representing the number when the function is finished. The third argument is a radix, which determines the number system in which your output is to be displayed. It must be between 2 and 36. The 10 used in the example gives you a decimal input. You could use 2 for binary, 8 for octal, 16 for hexadecimal, or some other number for your own weird number system. The function can return up to 17 bytes (including the null-terminating byte). This is the worst case situation when your radix is 2, representing a binary number, which is the longest one that can occur. Therefore, you should make sure your character array is at least 17 bytes long for the binary case or of some appropriate shorter length if you're using another radix. If you refer back to the discussion of the *printf* function, you'll see that it can display decimal, octal, or hexadecimal numbers. For binary numbers, or some other peculiar radix, you can use the *itoa* function and then use *printf* with the string printing format to display the resulting string.

The *ltoa* Function

The *ltoa* function is just the same as the *itoa* function except that it processes a number of data type *long* (integer). For this function, in the worst case 33 bytes can be returned to the string, so plan your character array size accordingly.

The *ultoa* Function

The *ultoa* function is just the same as the *itoa* function except that it processes a number of data type *unsigned long* (integer). For this function, in the worst case 33 bytes can be returned to the string, so plan your character array size accordingly.

The *ecvt* Function

The *ecvt* function converts a floating point number into a string. However, if you think you're going to use it in the same simple way as *itoa*, you're sadly mistaken. The *ecvt* function has a number of oddities in it that you have to be aware of and make provision for. Before continuing, let's look at a little program that makes use of the *ecvt* function.

```
/*

              ECONVERT = Test of ecvt function

                By Roger T. Stevens  8-13-92

*/
#include <stdio.h>
#include <stdlib.h>
#include <conio.h>
```

(continued)

```
void main(void)
{
        double n[10] = {-.3141596259, -3.141596259,
              -31415.96259, -3141596259.,
              .3141596259, 3.141596259, 31415.96259,
              3141596259., 314.159E4, 314.159E10};
        char *string;
        int decpoint, i, j, sign, ndigits=10, length;

        clrscr();
        for (j=0; j<10; j++)
        {
              string = ecvt(n[j], ndigits, &decpoint,
                    &sign);
              length = strlen(string);
              printf("\nstring = %s   decpoint = %d   sign"
                    " = %d",string, decpoint, sign);
              printf("\n");
              if (sign != 0)
                    printf("-");
              if (decpoint > 0)
                    for (i=0; i<decpoint; i++)
                          printf("%c",string[i]);
              else
                    decpoint = 0;
              printf(".");
              for (i=decpoint; i<length; i++)
                    printf("%c",string[i]);
        }
        getch();
}
```

The most important thing to note is that the *ecvt* function produces a string of numbers that does not include the sign or decimal point. The function is used like this

```
string = ecvt(n[j], ndigits, &decpoint, &sign);
```

where *string* contains the string of numbers returned by the *ecvt* function; the first argument, *n[j],* is the floating point number to be converted; the second argument, *ndigits*, is the number of digits to be contained in the string; and the next two arguments, *&decpoint* and *&sign*, are pointers to the parameters *decpoint* and *sign,* which will contain the decimal point location and sign information, respectively. You might think that you could define *string* as an array in the ordinary way

```
char string[45];
```

but this is not the case. You have to define *string* as a pointer.

Ordinarily if you did this and proceeded to use it without allocating any memory, you'd find yourself in big trouble, but in this case a static buffer area for *string* is built into the *ecvt* function. This buffer is overwritten each time *ecvt* is called. Now, if you're going to display the ASCII string generated by *ecvt* you'll need to make provision for inserting the decimal point. The argument *decpoint* shows the character location of the decimal point. If *decpoint* is negative or 0, the decimal point must be placed at the extreme left of the string of numbers. Otherwise, you display as many numbers as the value in *decpoint*, then insert the decimal point, and then display the rest of the numbers. You also need to provide for displaying the sign. If the argument *sign* contains a 0, the number is positive; otherwise the number is negative and you need to put a minus sign in front of it.

The value that you assign to *ndigits* is important. You need to have as many digits in your floating point number as you have assigned in *ndigits*. If you don't the function will pick up some garbage digits from somewhere and append them to the end of your actual number. You also have to be very careful when you convert a floating point number that is in the exponential format such as

```
a = 1.23689E6;
```

In this case, the string will contain as many characters as you have specified by *ndigits* and if the number is larger than this many digits, 0's will be appended at the right until a string of *ndigits* in length is produced. However, the decimal point location will be specified at the place where it should be and if this is further to the right than the number in *ndigits* the program above will have blank spaces between the last number in the string and the decimal point. You'll either need to make sure that such a situation doesn't occur or modify the program to insert 0's instead of spaces. The output of the program given is

```
string = 3141596259   decpoint = 0   sign = 1
-.3141596259
string = 3141596259   decpoint = 1   sign = 1
-3.141596259
string = 3141596259   decpoint = 5   sign = 1
-31415.96259
string = 3141596259   decpoint = 10   sign = 1
-3141596259.
string = 3141596259   decpoint = 0   sign = 0
.3141596259
string = 3141596259   decpoint = 1   sign = 0
```

```
3.141596259
string = 3141596259   decpoint = 5   sign = 0
31415.96259
string = 3141596259   decpoint = 10   sign = 0
3141596259.
string = 3141590000   decpoint = 7   sign = 0
3141590.000
string = 3141590000   decpoint = 13   sign = 0
3141590000
```

The *fcvt* Function

The *fcvt* function is exactly the same as the *ecvt* function except for the second argument, which specifies the number of digits. In the *ecvt* function, this argument specifies the total number of digits in the character string. In the *fcvt* function, this argument specifies the number of digits to the right of the decimal point.

The *gcvt* Function

The *gcvt* function is the one that you will want to use most of the time for floating point to ASCII character conversions, since it gives you a properly formatted string that will produce the number on your screen or printer without needing any manipulation. The form in which this function is used is

```
gcvt(value, ndigits, string);
```

where *value* is the floating point number to be converted. The argument *ndigits* is the number of significant digits in a standard floating point format, if this is possible to display; otherwise the number is displayed in exponential format (using powers of 10). The argument *string* is a pointer to the beginning of the character array where the result is to be stored. This is an ordinary character array. The following is a simple program that shows the use of *gcvt*.

```
/*

        GCONVERT = Test of gcvt function

          By Roger T. Stevens  8-13-92

*/
```

```
#include <stdio.h>
#include <stdlib.h>
#include <conio.h>

void main(void)
{
      double n[9] = {-.3141596259, -3.141596259,
            -31415.96259, -3141596259.,
            .3141596259, 3.141596259, 31415.96259,
            3141596259., 314.159E4};
      char string[25];
      int decpoint, i, j, sign, ndigits=10, length;

      clrscr();
      for (j=0; j<9; j++)
      {
            gcvt(n[j], ndigits, string);
            length = strlen(string);
            printf("\n%s",string);
      }
      getch();
}
```

Testing for Character Type

Before using some of the conversion functions that convert from ASCII strings to various types of data, you may want to make some tests to determine whether the string contains the right sort of data or not. The C language includes quite an assortment of such tests. You can determine whether a string contains numbers, letters, alphanumerics, etc., by just selecting the proper test. Each of the test functions determines whether a character is one of a particular character set, the character set differing for each individual test function. Table 15-I lists all of the test functions and their uses. One of these functions is used like this

```
k = isdigit(c);
```

where *c* is the character to be tested. After the function is called, the parameter *k* will contain a 0 if *c* is not in the character set and a nonzero value if *c* is in the character set. Since these functions only test one character at a time, you will need a loop to test a string, such as the following

Table 15-I. Character Testing Functions

Function	Description	Character Set
isalnum	Determines if a character is alphanumeric.	'A' to 'Z', 'a' to 'z', '0' to '9'
isalpha	Determines if a character is alphabetic.	'A' to 'A', 'a' to 'z'
isascii	Determines if a character is part of the ASCII character set.	0x00 to 0x7F
iscntrl	Determines if a character is a control character.	0x00 to 0x1F and 0x7F
isdigit	Determines if a character is a decimal digit.	'0' to '9'
isgraph	Determines if a character is printable (excluding the space).	0x21 to 0x7E
islower	Determines if a character is a lowercase (small) letter.	'a' to 'z'
isprint	Determines if a character is printable (including the space).	0x20 to 0x7E
ispunct	Determines if a character is a control character or a space.	0x00 to 0x20 and 0x7F
isspace	Determines if a character is a white space character.	0x09 to 0x0D and 0x20. In *printf* format these are: 0x09='\t', 0x0A='\n', 0x0B='\v', 0x0C='\f', 0x0D='\r', 0x20 = ' '
isupper	Determines if a character is an uppercase (capital) letter.	'A' to 'Z'
isxdigit	Determines whether a character is a hexadecimal digit.	'0' to '9', 'A' to 'F', 'a' to 'f'

```
char string[22];
int i;
...
...
i = 0;
while (string[i] != NULL)
{
   if ((isdigit(string[i])) == NULL)
      break;
   i++;
}
if (string[i] != NULL)
   printf("String contains a character other than 0-9");
else
   printf("String contains only characters 0-9");
```

If the string contains only decimal digits, the *while* loop will continue until the NULL character at the end of the string is encountered. When we exit from the loop, the character at *string[i]* will be that NULL character, giving an indication that the string is OK. If the string contains a character that is not a decimal digit, the program will break out of the loop at that point without checking the rest of the string. The character at *string[i]*, after leaving the loop, is the first character that is not a decimal digit, giving an indication that the string is not OK.

Newton's Method Fractals

Newton's method is a technique for finding a root of an equation by making a guess at the value of a root and then performing an iteration process, where each iteration approaches closer to the actual value of the root. Suppose that we have an equation

$$f(z) = 0 \qquad \text{(Equation 15-1)}$$

which has complex roots. The expression used for each iteration is

$$z_{n+1} = z_n + \frac{f(z_n)}{f'(z_n)} \qquad \text{(Equation 15-2)}$$

where $f'(z)$ is the derivative of the function. You'll observe that this iterated equation is similar to some of the iterated expressions used previously to produce fractal pictures. If you make a map of each display point in a section of the complex plane, with its color based on the root converged on from that point and the number of iterations needed to achieve a good approximation to the root, the resulting picture is a beautiful fractal. (You might think that the display would be dull, consisting of equal sections that converge on each root, but actually a selected point does not always converge on the nearest root, so the display is complicated and interesting.) The program listed produces a Newton's method display for solving the equation

$$z^3 - 1 = 0 \qquad \text{(Equation 15-3)}$$

This program is designed to show you examples of the use of *strtod* to convert strings of ASCII characters to integers. The function *strtod* is used several times at the beginning of the program to allow you to select the bounds for the region of the complex plane that is to be displayed, thereby expanding or contracting the fractal picture. The program to do this is listed in Figure 15-1 and the result of running this program with the parameters

```
Pmax = -0.210435
Pmin = -0.205171
Qmax = 0.050792
Qmin = 0.044194
```

is shown in Plate 13.

The program begins by displaying the legend *Enter new values if desired...* On the next line it displays *Maximum x:* followed by the default value for this parameter. The cursor returns to the first number of the default value, ready to replace it if desired. The program next enters a *do-while* loop, which continues to iterate until *end* (the pointer returned by *strtod*) is a NULL, indicating that a legitimate floating point number has been recorded. The program next enters a *for* loop, which permits you to enter up to 14 characters that are to comprise a floating point number. These characters are stored in an array called *number*. The loop ends when 14 characters have been entered or when the *Enter* key is hit. In this latter case, the 0x0D that is placed in the character string by the *Enter* key is replaced by a NULL to signify the end of the string. If the program exits the *for* loop by the *Enter* key being struck on the first character, the default value remains displayed. Otherwise, when the first character is entered, it replaces the first character of the default value and the remaining characters of the default value are blanked. They are then replaced by any additional characters that are typed in before the *Enter* key is struck. Next, the program checks whether the first character in the string is a NULL. If so, no changes are made to the default value of *Pmax*. This enables you to keep the default value by simply hitting the *Enter* key when you are presented with the default value on the screen. Next, the program calls *strtod* to convert the string in *number* to a floating point number, which is placed in *Pmax*. If the *end* argument is set to NULL, this means that the floating point conversion worked satisfactorily, so the program exits the *do-while* loop. If the value is not NULL, the program displays *Not a valid floating point number. Try again!* and then iterates through the loop again to give you another chance to enter a value for *Pmax*. If the next iteration produces a valid floating point number, the error message just given is blanked out and the program exits the *do-while* loop. The same procedure is then used to allow you to change the values of *Pmin, Qmax,* and *Qmin* if you so desire.

When the bounds have been established, the program computes the incremental change for each pixel in the *x* (*deltaP*) and *y* (*deltaQ*) directions. It then enters a pair of nested *for* loops that together iterate once for every pixel position on the screen. The inner loop begins by computing the initial *x* and *y* values for that pixel position (the starting guess of the root value for the Newton's method solution). The program then initializes the values of *xsq, ysq, old_x,* and *old_y*. Next, another *for* loop is entered, which iterates 64 times (at the maximum). Within this loop, the program uses the mathematics of Newton's method to compute the next approximation to the root. This is then compared with the value obtained from the previous iteration. When each component of the new root is within 1E-10 of the old one, the loop terminates. If this doesn't occur within 64 iterations, the loop terminates anyway. Of the 16 available colors, three groups of 5 are used to display pixel color. The group is selected by determining which of the three roots of the equation is being converged on. One of the five colors in the selected group is chosen by taking the number of iterations that occured modula 5. This process repeats until every pixel on the screen has been colored.

Figure 15-1. Listing of Program to Generate Newton's Method Fractals

```
/*

    NEWTON = Map of Newton's method for solving z^3 = 1

              By Roger T. Stevens  8-9-92

*/

#include <dos.h>
#include <stdio.h>
#include <math.h>
#include <stdlib.h>

int xres = 640, yres = 480;
int color, i, row, col;
char number[15], *end=number;
float Pmax=3.5, Pmin= -3.5, Qmax = 2.5, Qmin = -2.5;
double P, deltaP, deltaQ, denom, x, y, old_x, old_y, xsq,
      ysq, sqdif;

void plot(int x,int y,int color);
void setmode(int mode);
void cls(int color);

void main(void)
{
      setmode(3);
      gotoxy(10,10);
      printf("Enter new values if desired....");
      gotoxy(10,11);
      printf("Maximum x: %f",Pmax);
      do
      {
         gotoxy(21,11);
         for (i=0; i<14; i++)
         {
            number[i] = getche();
            if (number[i] == 0x0D)
            {
               number[i] = NULL;
               break;
            }
            if (i==0)
            {
               printf("                ");
               gotoxy(22,11);
            }
         }
```

(continued)

```
    if (number[0] != NULL)
    {
       Pmax = strtod(number, &end);
       gotoxy(10,15);
       if (*end != NULL)
          printf("Not a valid floating point number."
             "Try again!");
       else
          printf("                                                   "
             "             ");

    }
 }
 while (*end != NULL);
 gotoxy(10,12);
 printf("Minimum x: %f",Pmin);
 do
 {
    gotoxy(21,12);
    for (i=0; i<14; i++)
    {
       number[i] = getche();
       if (number[i] == 0x0D)
       {
          number[i] = NULL;
          break;
       }
       if (i==0)
       {
          printf("                ");
          gotoxy(22,12);
       }
    }
    if (number[0] != NULL)
    {
       Pmin = strtod(number, &end);
       gotoxy(10,15);
       if (*end != NULL)
          printf("Not a valid floating point number."
             " Try again!");
       else
          printf("                                                   "
             "             ");

    }
 }
 while (*end != NULL);
 gotoxy(10,13);
 printf("Maximum y: %f",Qmax);
 do
 {
    gotoxy(21,13);
    for (i=0; i<14; i++)
```

(continued)

```
        {
           number[i] = getche();
           if (number[i] == 0x0D)
           {
              number[i] = NULL;
              break;
           }
           if (i==0)
           {
              printf("                ");
              gotoxy(22,13);
           }
        }
        if (number[0] != NULL)
        {
           Qmax = strtod(number, &end);
           gotoxy(10,15);
           if (*end != NULL)
              printf("Not a valid floating point number."
                 " Try again!");
           else
              printf("                                    "
                 "            ");

        }
     }
     while (*end != NULL);
     gotoxy(10,14);
     printf("Mimimum y: %f",Qmin);
     do
     {
        gotoxy(21,14);
        for (i=0; i<14; i++)
        {
           number[i] = getche();
           if (number[i] == 0x0D)
           {
              number[i] = NULL;
              break;
           }
           if (i==0)
           {
              printf("                ");
              gotoxy(22,14);
           }
        }
        if (number[0] != NULL)
        {
           Qmin = strtod(number, &end);
           gotoxy(10,15);
           if (*end != NULL)
              printf("Not a valid floating point number."
                 " Try again!");
```

(continued)

```
            else
               printf("                                     "
                  "           ");
         }
      }
      while (*end != NULL);
      setmode(18);
      cls(7);
      deltaP = (Pmax - Pmin)/xres;
      deltaQ = (Qmax - Qmin)/yres;
      for(col=0; col<xres; col++)
      {
         for(row=0; row<yres; row++)
         {
            x = Pmin + col*deltaP;
            y = Qmax - row*deltaQ;
            xsq = ysq = 0;
            old_x = old_y = 42;
            for (i=0; i<64; i++)
            {
               sqdif = (x - y) * (x + y);
               xsq = x*x;
               ysq = y*y;
               denom = 3.0*(sqdif*sqdif + 4.0*xsq*ysq);
               if (denom == 0)
                  denom = 1E-10;
               y = 0.6666667*y - 2.0*x*y/denom;
               x = 0.6666667*x + sqdif/denom;
               if (fabs(old_x - x) < 1E-10 && (fabs(old_y -
                  y) < 1E-10))
                  break;
               old_x = x;
               old_y = y;
            }
            if (x>0)
               color = i%5;
            else
               if ((x < -0.3) && (y>0))
                  color = (i%5) + 5;
               else
                  color = (i%6) + 10;
            plot(col, row, color);
         }
      }
      getch();
}

/*

                    setmode() = Sets video mode

 */
```

(continued)

```
void setmode(int mode)
{

        union REGS reg;

        reg.x.ax = mode;
        int86 (0x10,&reg,&reg);
}

/*
                cls() = Clears the screen
*/

void cls(int color)
{
        union REGS reg;

        reg.x.ax = 0x0600;
        reg.x.cx = 0;
        reg.x.dx = 0x1E4F;
        reg.h.bh = color;
        int86(0x10,&reg,&reg);
}

/*
   plot() = Plots a point on the screen at a designated
   position using a selected color for 16 color modes.
*/

void plot(int x, int y, int color)
{
        #define graph_out(index,val)  {outp(0x3CE,index);\
                                outp(0x3CF,val);}

        int dummy,mask;
        char far * address;

        address = (char far *) 0xA0000000L + (long)y *
           xres/8L + ((long)x / 8L);
        mask = 0x80 >> (x % 8);
        graph_out(8,mask);
        graph_out(5,2);
        dummy = *address;
        *address = color;
        graph_out(5,0);
        graph_out(8,0xFF);
}
```

16

Manipulating Strings

The C language contains a wide variety of functions for manipulating strings of characters. You'll probably never use all of them, but as you're working on a program, you'll probably encounter a difficult problem with strings that will turn out to have a perfect solution by using one of the string functions. Since you never can be sure which string function is the one that is the answer to your problem, you need to have some familiarity with what each of these functions can do. Most documentation that comes with C compilers simply lists and describes each string function in alphabetical order. In this chapter, we've grouped the functions according to the operations they perform so that if you have a problem, you can go to the proper table and find all of the functions that will do what you want and examine the differences to find the exact function that is a perfect match for the problem.

String Copying Functions

These functions copy the contents of an existing string to another. The functions *stpcpy* and *strcpy* are the same. They are listed in Table 16-1. The functions require, as arguments, the address of an existing destination character array and the address of an existing source character array which contains the string to be copied. The function *strdup* requires a pointer to a destination array that does not yet exist and the address of a source array containing the string to be copied. The function uses *malloc* to assign space for the destination array and performs the copying function. The function returns the address of the destination character array, or a NULL if memory could not be

Table 16-I. String Copying Functions

Function	Description	Usage
stpcpy	Copies the contents of a source string to a destination string.	stpcpy(destination, source);
strcpy	Copies the contents of a source string to a destination string. (Same as *stpcpy*.)	stpcpy(destination, source);
strdup	Copies a string to a newly created location. Makes duplicate string, obtaining memory space with *malloc*.	char * new_string; ... new_string = strdup(old_string);
strncpy	Copies a designated number of bytes (*no_of_characters*) from source string to destination string. If the source string doesn't have enough characters, the destination will be padded with NULLS. If source string is longer than the number of bytes specified, the destination string may not be NULL terminated.	strncpy(destination, source, no_of_characters);
strxfrm	Copies a designated number of bytes (*no_of_characters)* from the source to the destination string. If the source string doesn't have enough characters, the destination will be padded with NULLS. If the source string is longer than the number of bytes specified, the destination string may not be NULL terminated. Returns length of destination string.	length = strncpy(destination, source, no_of_characters);

assigned to it. The function *strncpy* takes as arguments the address of a destination character array, the address of a source character containing the string to be copied, and the number of bytes to be copied. This number of bytes will be array transferred. If there are not enough bytes, the destination string will be padded with NULLS; if there are too many, the destination string may not have a terminating NULL. The function *strxfrm* is like *strncpy* except that it returns the length of the destination string.

String Appending Functions

These functions append the contents of one existing string to another. The function *strcat* requires as arguments the address of an existing destination character array and the address of an existing source character array. The contents of the source array are appended at the end of the existing destination string. The function *strncat* requires as arguments the address of an existing destination character array and the address of an existing source character array, and an integer representing the number of bytes (characters) to be transferred. This number of bytes from the beginning of the source array are appended at the end of the existing destination string. These functions are listed in Table 16-II.

Table 16-II. String Appending Functions

Function	Description	Usage
strcat	Appends the contents of a source string to a destination string.	strcat(destination, source);
strncat	Appends, at most, a designated number of bytes *(no_of_characters)* from the source string to the destination string, and then appends a NULL character.	strncat(destination, source, no_of_characters);

String Comparison Functions

These functions perform various comparisons of two strings. The function *strcmp* compares two strings and returns a negative number if the first string is smaller than the second, a 0 if the two strings are the same, and a positive number if the first string is larger than the second. The functions *stricmp* and *strcmpi* are identical. They compare two strings in the same way as just described and with the same returns, except that the comparison is made without case sensitivity (in other words no distinction is made between capitals and small letters). The function *strncmp* is the same as *strcmp* except that it requires a third argument, namely the number of bytes (characters) to be compared. It only compares this number of characters from the beginning of each of the two strings. The functions *strncmpi* and *strnicmp* are identical and are the same as *strncmp* except that they make the comparison without case sensitivity. See Table 16-III.

String Character Search Functions

These functions scan a string to find the occurrence of specified characters. The function *strchr* scans a string from the beginning for the first occurrence of a specified character. It returns a pointer to this character. The function *strpbrk* returns a pointer to the first occurrence in a string of any character from a set of characters given in a test string. The function *strrchr* is the same as *strchr* except that it searches the string in the reverse direction, so that it returns a pointer to the last occurrence of the character. See Table 16-IV.

String Subset of Characters Search Functions

These functions look for the presence or absence of a subset of characters in a string. The function *strcspn* scans a string and returns the length from the beginning of the string to the first occurrence of any one of the characters in *test_string*. The function *strspn* is similar, except that it returns the length of the portion of the string to the first character that is not contained in *test_string*. The function *strstr* scans a string for the first occurrence of the entire substring defined by *sub_string* and returns a pointer to the location of the beginning of this substring. See Table 16-V.

Table 16-III. String Comparison Functions

Function	Description	Usage
strcmp	Compares *string_1* with *string_2*. Returns: <0 if *string_1* < *string_2* =0 if *string_1* = *string_2* >0 if *string_1* > *string_2*.	strcmp(string_1, string_2);
strcmpi	Compares *string_1* with *string_2* without case sensitivity. Returns: <0 if *string_1* < *string_2* =0 if *string_1* = *string_2* >0 if *string_1* > *string_2*.	strcmpi(string_1, string_2);
stricmp	Same as *strcmpi*.	stricmp(string_1, string_2);
strncmp	Compares a designated number of bytes (*no_of_characters*) of *string_1* with *string_2*. Returns: <0 if *string_1* < *string_2* =0 if *string_1* = *string_2* >0 if *string_1* > *string_2*.	strncmp(string_1, string_2, no_of_characters);
strncmpi	Compares a designated number of bytes (*no_of_characters*) of *string_1* with *string_2* without case sensitivity. Returns: <0 if *string_1* < *string_2* =0 if *string_1* = *string_2* >0 if *string_1* > *string_2*.	strncmpi(string_1, string_2, no_of_characters);
strnicmp	Same as *strncmpi*.	strnicmp(string_1, string_2, no_of_characters);

Table 16-IV. String Character Search Functions

Function	Description	Usage
strchr	Scans string for the first occurrence of a specified character. Returns pointer to string member where first occurrence of character took place.	char string[45]; char character; char *pointer; ... ... pointer = strchr(string, character);
strpbrk	Scans a string for the first occurrence of a character from a given set.	char string[45]; char test_string[34]; char *pointer; ... ... pointer = strpbrk(string, test_string);
strrchr	Scans string in the reverse direction looking for the last occurrence of a specified character. Returns pointer to string member where last occurrence of character took place.	char string[45]; char character; char *pointer; ... ... pointer = strrchr(string, character);

Uppercase/Lowercase Conversions

The function *strlwr* converts capitals to small letters. The function *strupr* converts small letters to capitals. No other characters are changed. The functions are listed in Table 16-VI.

Table 16-V. String Subset of Characters Search Functions.

Function	Description	Usage
strcspn	Scans string *string* to determine initial segment not containing any of a set of characters specified by *test_string*. Returns length of segment not containing any of the characters in *test_string*.	char string[45]; char test_string[15]; int length; ... ... length = strcspn (string, test_string);
strspn	Scans string *string* to determine initial segment containing nothing but characters from a set specified by *test_string*. Returns length of this segment.	char string[45]; char test_string[15]; int length; ... ... length = strspn (string, test_string);
strstr	Scans string *string* for the first occurrence of a substring given by *sub_string*.	char string[45]; char sub_string[15]; char *pointer; ... ... pointer = strstr(string, sub_string);

Setting Characters in a String

The function *strset* sets all characters in a string to a designated character and the function *strnset* sets a specified number of characters to a designated character. At first glance, one might think that if a character array was defined by

```
char array[40];
```

one of these functions could be used to initialize all or part of the array to a desired character. Unfortunately, this is not true, because both functions stop when they encounter a NULL (to indicate the end of the string) which is the first character in the array shown. Thus you need to use the *memset* function described in Chapter 17 to do this kind of initialization. These functions are listed in Table 16-VII.

Converting Time or Date to a String

The function *_strdate* obtains the current date from the system clock and converts it to a string of form *MM/DD/YY*, each of these being a two-digit number representing the month, day, and year respectively. The function *_strtime* obtains the current time from the system clock and converts it to a string of form *HH:MM:SS*, each of these being a two-digit number representing the hour, minute, and second, respectively. The function *strftime* is similar to the *printf* family of functions, but it is designed to handle time variables. See Table 16-VIII. The functions to handle time and date will be described in detail in Chapter 18. This function, like others that work with time and date, requires some peculiar operations. In addition, it stores the string that it creates in a character array that you have to display with a *printf* statement.

Table 16-VI. String Uppercase/Lowercase Conversion Functions

Function	Description	Usage
strlwr	Converts uppercase characters (A to Z) in string *string* to lowercase characters (a to z). No other characters are changed. Returns a pointer to the string.	char string[45]; ... ... strlwr (string);
strupr	Converts lowercase characters (a to z) in string *string* to uppercase characters (A to Z). No other characters are changed. Returns a pointer to the string.	char string[45]; ... ... strupr (string);

Table 16-VII. Functions to Set Characters in String

Function	Description	Usage
strnset	Sets a specified number of characters (*no_of_characters*) in a string (*string*) to a designated character (*character*). Stops when a NULL is encountered, even if *no_of_characters* has not been reached, so it cannot be used to initialize an empty buffer.	char string[45]; char character; int no_of_characters ... ... strnset (string, character, no_of_characters);
strset	Sets all characters in a string (*string*) to a designated character (*character*). Stops when a NULL is encountered, so it cannot be used to initialize an empty buffer.	char string[45]; char character; ... ... strset (string, character);

Here is a program that shows how to use *strftime*

```
#include <time.h>
#include <stdio.h>

void main(void)
{
      struct tm *tblock;
      time_t timer;
      char string[80];

      clrscr();
      time (&timer);
      tblock = localtime(&timer);
      strftime(string,80,"Today is %A, %B %d, %Y. The"
            time is %I:%M %p.",tblock);
      printf("\n%s\n",string);
      getch();
}
```

Table 16-VIII. Functions to Convert Time or Date to String

Function	Description	Usage
_strdate	Converts current date (of system clock) to a string (*string*). The string must be at least nine characters long. The date is stored in the form *MM/DD/YY*, where *MM*, *DD*, and *YY* are two-digit numbers representing the month, day, and year, respectively.	char string[9]; ... ... _strdate(string);
strftime	Formats and displays time in a similar way to *printf*. See description for details.	See description for details.
_strtime	Converts current time (of system clock) to a string (*string*). The string must be at least nine characters long. The date is stored in the form *HH:MM:SS*, where *HH*, *MM* and, *SS* are two-digit numbers representing the hour, minute, and second, respectively.	char string[9]; ... ... _strtime(string);

The program first uses *time* to get the elapsed time in seconds since January 1, 1970, and store it in *timer*. It then uses the function *localtime* to break this time up into years, months, days, etc., and store the new information in the structure *tblock*. Next *strftime* builds a string of time data using format directives from the list in Table 16-IX. These directives are similar to those used with the *printf* family, but they do not require individual arguments to supply values. Instead only one argument is necessary, the structure *tblock*. (Of course you can name it what you want as long as it is a structure of type *tm*. This structure will be described in Chapter 18.) The string is stored in *string* from whence it is printed by *printf*.

Table 16-IX. Format Directives for strftime Function

Format Directive	Replaced by
%%	character %
%a	abbreviated weekday name
%A	full weekday name
%b	abbreviated month name
%B	full month name
%c	date and time in the form *Wed Aug 19 14:32:12 1992*
%d	two-digit day of the month (01 to 31)
%H	two-digit hour (00 to 23) for 24 hour clock
%I	two-digit hour (01 to 12) for 12 hour clock
%j	three-digit day of the year (001 to 366)
%m	two-digit month (01 to 12)
%M	two-digit minute (00 to 59)
%p	AM or PM
%S	two-digit second (00 to 59)
%U	two-digit week no. (00 to 53) (1st day is Sunday)
%w	weekday (0 to 6) (Sunday is 0)
%W	two-digit week no. (00 to 53) (1st day is Monday)
%x	date (in form *Wed Aug 19, 1992*)
%X	time (in form *14:45:32*)
%y	two-digit year (00 to 99) (doesn't include century)
%Y	four-digit year (0000 to 9999) (including century)
%Z	time zone name (Null string if no time zone)

The *strlen* Function

You will sometimes need to know the length of a string. Probably you can figure out a simple little function to find this for you by using a *for* loop to look at each character in the string and stop when a NULL is encountered. Fortunately, however, the C library contains a function that will do the job for you. It is called *strlen*. It accepts one argument, the address of a character array containing a string, and returns the length of the string. It is described in Table 16-X.

Table 16-X. String Length Function

Function	Description	Usage
strlen	Returns the length of a string (*string*).	char string[56]; int length; ... ... length = strlen(string);

Converting Strings to Numbers

This group of functions converts strings of characters to numbers. The function *strtod* converts a string to a floating point number of type *double*. The function *strol* converts a string to an integer of type *long* integer. The function *stroul* converts a string to an integer of type *unsigned long*. Each of these functions requires a string with a particular format for a legitimate conversion to take place. The formats are included in the descriptions in Table 16-XI. Each function also has an argument that can be used to determine whether the conversion was made successfully, and, if not, which character caused the conversion to fail.

Table 16-XI. Functions to Convert Strings to Numbers

Function	Description	Usage
strtod	Converts string (*string*) to a number (*result*) of type *double*. String must have format of *[ws] [sn] [ddd] [.] [ddd] [e]* or *[E] [sn] [ddd]* where *ws* = white space, *sn* = sign, *[ddd]* = digits. The argument *endptr* is NULL if the conversion was OK; otherwise it points to the character that stopped the conversion.	char string[80]; double result; char *endptr; ... ... result = strtod(string, &endptr);
strtol	Converts string (*string*) to a number (*result*) of type *long*. String must have format of *[ws] [sn] [0] [x]* or *[X] [ddd]*, where *ws* = white space, *sn* = sign, *[ddd]* = digits. The argument *endptr* is NULL if the conversion was OK; otherwise it points to the character that stopped the conversion.	char string[80]; long result; char *endptr; ... ... result = strtol(string, &endptr);
strtoul	Converts string (*string*) to a number (*result*) of type *unsigned long*. String must have the format of *[ws] [sn] [0] [x]* or *[X] [ddd]*, where *ws* = white space, *sn* = sign, *[ddd]* = digits. The argument *endptr* is NULL if the conversion was OK; otherwise it points to the character that stopped the conversion.	char string[80]; unsigned long result; char *endptr; ... ... result = strtoul(string, &endptr);

Reversing a String

The function *strrev*, described in Table 16-XII, reverses the contents of a string, except for the terminating NULL character which remains at the end of the reversed string.

Table 16-XII. String Reverse Function.

Function	Description	Usage
strrev	Reverses a string, except for the terminating NULL character. For example, *string\0* would become *gnirts\0*.	char string[56]; ... ... strrev(string);

String Token Search

The function *strtok*, which is described in Table 16-XIII, permits you to define a delimiter which separates different parts of a string from each other. Using this function you can then separate these parts. The following program provides an illustration:

```
#include <stdio.h>
#include <string.h>
#include <conio.h>

char data[64] =
      {"Roger T. Stevens*9717 Regal Ridge"
      " NE*Albuquerque, NM 87111"};
char *pointer;

void main(void)
{
      printf("\n%s\n\n",data);
      pointer = strtok(data,"*");
      if (pointer)
            printf("%s\n",pointer);
      pointer = strtok(NULL,"*");
      if (pointer)
            printf("%s\n",pointer);
      pointer = strtok(NULL,"*");
      if (pointer)
            printf("%s\n",pointer);
```

(continued)

```
        printf("\n%s",data);
        getch();
    }
```

You begin by defining a pointer (*pointer* in the example). You set *pointer* equal to the return from the function *strtok*, which has two arguments, the address of the file containing the data to be processed (*data*) and the delimiter character or characters (*). A NULL has now been inserted in place of the first delimiter and the pointer points to the beginning of the string. The following *printf* statement then prints out the section of the data from the beginning to the delimiter. From now on, repeated calls to *strtok* for this piece of data have a NULL as the first argument. This causes the pointer to point to the beginning of the part of the string after the first delimiter and replace the second delimiter by a NULL. The next *printf* statement then prints out this next section of the string. This process can be continued for each section that makes up the string. Be warned, however, that the string is permanently altered by calling the *strtok* function. The original delimiters are gone, so that if you were to attempt to run this program twice on the same string, it would not work the second time. If you must do repeated *strtok* operations on the same data string, you should first copy it to a temporary location and then operate on the copy so that the original is not changed.

Table 16-XIII. String Token Search Function.

Function	Description	Usage
strtok	Searches a string for tokens which are separated by delimiters defined in a second string.	char string[56]= "abc,def,ghi"; char * p; ... ... p = strtok(string, ",");

The *readfile.c* Program

Back in Chapter 10, we generated the fractal curve for the Tchebychev C_5 polynomial and saved it to disk as *tcheb.raw*. We also had a program to read this file from disk and display it on the screen. In Chapter 11, we revisited the file reading problem, with a program that instead of having a fixed file name to load could load any file whose name was inserted with the keyboard and read into the program using the *scanf* function. This didn't provide much error protection, however. The program that follows in Figure 16-1 uses various of the string manipulating functions to read in a file name from the keyboard and check it for every possible kind of error. The program requires that you already have the file *tcheb.raw*, or a file with the same structure in the same directory as this program.

There are five different kinds of errors that must be provided against. These are:

(1) Illegal characters in file name. DOS will allow only certain characters to be used as part of a file name.

(2) Too many characters in the file name. DOS allows only eight characters before the period that marks the boundary between the file name proper and the extension.

(3) Too many characters in the extension. DOS allows only three characters in the extension.

(4) Too many periods in the file name. DOS allows only one.

(5) File of this name not on disk. This implies that the file name complied with all of the DOS requirements, but when we went to open the file, a file of this name could not be found on the disk.

In addition, if the file name does not include an extension, we want to add the period and the extension *RAW*. The program begins by asking you to enter the file name. It then enters a most peculiar *while* loop. This loop simply contains a 1 as the test condition. Ordinarily, any test that is defined by a conditional statement returns a 1 for *True*, causing the loop to continue to iterate, or a 0 for *False*, causing the loop to come to an end. By just using a 1, we set up a condition that is always true, so the *while* loop will loop forever. After initializing a couple of variables, the program enters another *while* loop, which allows you to enter up to 12 characters to comprise a file name. Each is put into the *filename* buffer. If the character is a carriage return, this inner loop terminates. Otherwise, it continues to the last permissible character. In addition, if a period is entered in the file name string, its location is stored in *period*. Each character is also displayed on the screen. After the loop ends, the string is terminated with a NULL character. Next, the function *strlen* is used to find the length of the file name string. If *period* was 0, indicating that no period was included in the file name, the location of period is set to the value in *length*. (For no period in the name, the implied period is at the end of the file name string.) Next we use the function *strcspn* to search the file name for the length of the segment of the string that does not contain any character that is illegal for a DOS file name. All of these illegal characters are in the string called *teststring*. If this length is the same as the length of the file name, there is no illegal character present and we continue. Otherwise we set the argument *error* to 1. Next we check whether the period position indicates that it followed more than eight characters. If so, we set *error* to 2. Otherwise, we use *strchr* to scan for the first occurrence of a period. If it returns a NULL, indicating no period, we check whether the length of the file name (without extension) is greater than eight characters. If so, we set *error* to 2. Next, we check whether the string length *(length)* is greater than the period position plus 4. If so, there are more than the three allowable extension characters, so *error* is set to 3. Finally, we use *strchr* to scan the file name forward for a period and then *strrchr* to scan the file name string backwards for a period. If the location for both of these is the same, there is only one period in the file name; otherwise, there is more than one period and we set *error* to 4.

Now, if *error* is still 0, none of the errors has occurred and we have a legally correct file name. If this is the case, we check using *strchr* whether there is a period in the name; if not, we use *strcat* to append the period and the extension *RAW*. Then, we try to open the file having this file name. If *fsave* has a pointer in it at this point, we have opened the file satisfactorily so we break from the infinite *while* loop and continue with the rest of the program. If *fsave* is left with a NULL, the system cannot find the file on the disk, so we set *error* to 5. We then blank out the location where an error message would appear and then write the proper error message there for the new value of *error*. Below it we write *Try again!*. Another pass is then made through the infinite *while* loop, and this continues until the program is able to open a legitimate file.

The rest of the program sets the display to the graphics mode and transfers the data from disk to the display as was explained for the two earlier versions of the program.

Figure 16-1. Program to Read and Display a Graphics File

```
/*

        READFILE = Program to read a graphics information
              disk file and display it on the screen
               with protection for getting correct
                          file name.

                  By Roger T. Stevens  8-12-92

*/

#include <stdio.h>
#include <dos.h>
#include <conio.h>

void setmode(int mode);

FILE *fsave;
char buffer[38400];
int i=0, length, period, error;
char teststring[128] =
      {0x01,0x02,0x03,0x04,0x05,0x06,0x07,0x08,0x09,0x0A,
      0x0B,0x0C,0x0D,0x0E,0x0F,0x10,0x12,0x13,0x14,0x15,
      0x16,0x17,0x18,0x19,0x1A,0x1B,0x1C,0x1D,0x1E,0x1F,
      0x20,0x22,0x2A,0x2B,0x2C,0x3A,0x3B,0x3C,0x3D,0x3E,
      0x3F,0x5B,0x5C,0x5D,0x7C,0x7F};
char filename[14];
char errormsg[6][48] = {{NULL},
      {"Illegal character in file name "},
      {"Only 8 characters allowed in file name "},
      {"Only 3 characters allowed in file extension "},
      {"Only one period allowed in file name"},
      {"Unable to find file "}};

void main(void)
{
      setmode(3);
      gotoxy(10,10);
      printf("Enter file name: ");
      while(1)
      {
            i = 0;
            error = 0;
            while (i < 13)
```

(continued)

```
        {
              filename[i] = getch();
              if (filename[i] == 0x0D)
                    break;
              if (filename[i] == 0x2e)
                    period = i;
              printf("%c", filename[i++]);
        }
        filename[i] = NULL;
        length = strlen(filename);
        if (period == 0)
              period = length;
        if (length != strcspn(filename, teststring))
              error = 1;
        if ((period > 8) || ((strchr(filename,0x2E) ==
              NULL) && (length > 8)))
              error = 2;
        if (length > period + 4)
              error = 3;
        if (strchr(filename,0x2e) !=
              strrchr(filename,0x2E))
              error = 4;
        if (error == 0)
        {
              if (strchr(filename,0x2e) == NULL)
                    strcat(filename,".RAW");
              fsave = fopen(filename,"rb");
              if (fsave != NULL)
                    break;
              error = 5;
        }
        gotoxy(10,10);
        printf("%s '%s'                              ",
              errormsg[error], filename);
        gotoxy(10,11);
        printf("Try again!                           ");
        gotoxy(21,11);
  }
  setmode(18);
  for (i=0; i<4; i++)
  {
        fread(buffer,1,38400,fsave);
        outport(0x3C4,(0x01<<(8+i)) + 2);
        movedata(FP_SEG(buffer),FP_OFF(buffer),
              0xA000,0,38400);
  }
  fclose(fsave);
  getch();
}
```

(continued)

```
/*
┌──────────────────────────────────────────────┐
│                                              │
│          setmode() = Sets video mode         │
│                                              │
└──────────────────────────────────────────────┘
*/

void setmode(int mode)
{

      union REGS reg;

      reg.x.ax = mode;
      int86 (0x10,&reg,&reg);
}
```

17

Memory Management

As long as you don't have any very large arrays and as long as you don't have an excessive number of variables, the techniques described thus far in this book will enable you to write successful C programs without any difficulty. All of the programs described so far work just fine with the small memory model, which produces the fastest and most compact programs. However, at some point, you may find that you need to handle large quantities of data and the simple techniques just don't work any more. This chapter will describe how to handle large amounts of data and will show you some of the pitfalls that occur when you try to do this. Working with a lot of data introduces a number of additional ways in which you can err in your C programs. Some of the errors are very subtle; you'll see very strange things on your screen that don't appear to make much sense. Therefore you really need to be on your toes in this situation.

Memory Models

Most C compilers for use with the IBM PC and compatibles have a number of memory models that can be used. Selection of a memory model determines how much memory can be allocated for the program and for the data. The memory models come about because of the segmented nature of PC memory. If normal address pointers are used, the program being written can only access 64K of memory. This limits how large a program or how much data you can have, but it also enables the compiler to use the shortest and fastest memory address pointers. When you select a memory model that uses far memory address pointers, you can access larger chunks of memory, but the compiler has to use longer addresses, which enlarges the program and makes it run slower. The memory models available with most C compilers are listed in Table 17-I. The table gives you some idea of how to match the memory model to the size of your code and data. The memory model is selected in Borland C++ by typing Alt-O to choose *Options* and then selecting *Compiler* and then *Code Generation*. The resulting menu gives you an opportunity to choose your memory model. The memory model selection goes into effect when you next compile your program listing. Since no changes are required in the program listing to use a larger memory model, you can usually start with the *Small* memory model and switch to a larger one if you encounter difficulty in compiling.

Things You Can't Do

Take a look at this small program:

```
#include <stdio.h>

char r[66000];

void main(void)
{
        r[4] = 'r';
        r[65800] = 's';
        printf("%c %c", r[4], r[65800]);
        getch();
}
```

If you try to compile it, you will get an error message that says *Array size too large*. This is because no array in C on the PC can be larger than 64K (65,536) bytes. Now how about this program:

```
#include <stdio.h>

char r[33000], s[33000];

void main(void)
{
        r[4] = 'r';
        r[32900] = 's';
        printf("%c %c", r[4], r[32900]);
        getch();
}
```

If you attempt to compile this program, you'll get an error message that states *Too much global data defined in file*. This is true no matter what memory model you use. Even with the *huge* memory model, you cannot have more that 64K of global memory. Here's a

Table 17-1. Memory Models for C Compilers

Model	**Code Size**	**Data Size**	**Use**
Tiny	64KB		One 64K segment for both code and data. Use for smallest programs and for generating *.com* files.
Small	64KB	64KB	Code and data segments don't overlap. Good for average applications.
Medium	1MB	64KB	Far pointers used for code but not data. Good for large programs without much data.
Compact	64KB	1MB	Far pointers used for data but not code. Good for small programs with a lot of data.
Large	1MB	1MB	Far pointers for everything. Static data limited to 64K. For very large applications.
Huge	1MB	1MB	Far pointers for everything. 1 MB static data limit. For very large applications.

third example program

```
#include <stdio.h>

void func(void);

void main(void)
{
        static char r[33000];

        r[4] = 'r';
        r[32900] = 's';
        printf("%c %c", r[4], r[32900]);
        func();
        getch();
}

void func(void)
{
        static char s[33000];
}
```

You might think that you could compile this program in the *huge* memory model, since that model allows you to have more than 64K of static data. Unfortunately, when you attempt to compile, you get the same old error message *Too much global data defined in file*. This is because even with the *huge* memory model, you cannot have more than 64K of static data within a module. If you placed the function *func* in a separate module and compiled it separately, you could then link it together with the *main* function without any error messages, provided you used the *huge* memory model.

The *malloc* and *calloc* Functions

Suppose we have a program that, in part, looks like this

```
#include <stdio.h>

...
...

void func(void);

void main(void)
{
        ...
        ...
```

(continued)

```
        ...
        func();
}

void func(void)
{
        char r[48000]
        r[4] = 'r';
        r[47000] = 's';
        printf("%c %c", r[4], r[47000]);
        getch();
}
```

What is going to happen when you compile and run this program? The answer is that it all depends on the part of the program that is not shown. If you have too much data in that unshown part of the program, you may get an error message when you compile. Worse yet, the program may compile all right, but there may not really be enough memory available for the array *r*. When the program runs, the value that you attempt to insert into *r[47000]* may actually overlap some other part of the program in memory. It's anybody's guess what the program will do after you insert this value into memory. You may just change some data for another argument, so that somewhere in the program you may find an incorrect data item. However, you may also change something that is part of the program operating structure, so that the program goes off and does something completely unexpected. (This might be something awful, like erasing part of your hard disk.)

The functions *malloc* and *calloc* provide a vehicle for assigning memory to a large array and making sure that such memory is actually available. Here is a good way to use the *malloc* function

```
char * bigarray;
...
...
if ((bigarray = malloc(33000)) == NULL)
{
        printf("\nNot enough memory for bigarray...");
        getch();
        exit(0);
}
```

Near the top of your program or function, you define *bigarray* as a pointer to data of type *char*. Then, at some point before you are first going to make use of the *bigarray* array, you need to allocate memory for it. This is done by the statement

```
bigarray = malloc(33000)
```

This is imbedded in an *if* statement that tests whether a NULL was returned to the *bigarray* pointer. Upon returning from *malloc, bigarray* should contain the address of the block of memory that has been assigned to the array. If the program was unable to find enough memory for the array, it returns a NULL instead of an address pointer in *bigarray*. The section of code given above returns an error message stating that memory isn't available when this condition occurs. The program then calls *getch* to wait for the operator to acknowledge by entering a keystroke after which it terminates the program with the *exit* statement. If you don't use the *getch* statement, the program may zip by the error message so fast that you don't see it. One big advantage of defining a big array in this way is that you always know if the program continues normally that there is enough memory for the array so that no strange things are going to occur because of memory overlap. Once the memory has been allocated, you can refer to elements of the array just as if it were an ordinary array, with statements such as

```
k = bigarray[28970];
```

In the example, we defined *bigarray* to be a pointer to an array of characters. It's possible to define the array to be of any data type. For example, you could have

```
int * bigarray;
...
...
if ((bigarray = malloc(33000)) == NULL)
{
        printf("\nNot enough memory for bigarray...");
        getch();
        exit(0);
}
```

The thing to note here is that we have defined *bigarray* to be a pointer to an array of integers. When we call *malloc*, we specify the number of **bytes** of memory that are observed for the array. Since each integer occupies two bytes, if we refer to a member of the array by

```
k = bigarray[3360];
```

then we are getting the contents of the 6720 and 6721 bytes of the array. Thus the 16499th member of the array is the last one. It occupies bytes 33498 and 33499. If you attempt to use any larger number for the array member, you will overlap some other portion of memory and set up for a potential disaster. (The C compiler won't tell you if you do this, it will compile and run as if everything were OK until the memory overlap causes your program to go haywire.)

The *calloc* function is very similar to the *malloc* function except that it zeroes the memory block assigned to the array. This slows things down, so use *malloc* except when you really need to have the memory block zeroed. The *calloc* function also has a second argument that is passed to it, which is the size of each data item in the array. Thus you could have

```
char * bigarray;
...
...
if ((bigarray = calloc(33000,1)) == NULL)
{
        printf("\nNot enough memory for bigarray...");
        getch();
        exit(0);
}
```

where the call to *calloc* states that there are 33000 data items of size 1 byte, which results in 33000 bytes being assigned to the array. This looks almost like what we did with *malloc*. However in the integer data type case, things are a little different. We have

```
int * bigarray;
...
...
if ((bigarray = calloc(16500,2)) == NULL)
{
        printf("\nNot enough memory for bigarray...");
        getch();
        exit(0);
}
```

Here the call to *calloc* says that there are 16500 data items, each having a length of 2 bytes, so that the actual memory allocated to the array is 33000 bytes. This is really a little easier to understand than is the use of *malloc*.

The Stack and the Heap

Most C compilers divide the available memory into two areas, the *stack* and the *heap*. All variables that are defined in the main program and in the functions are assigned to memory space on the stack by the compiler. The compiler begins assigning space on the stack at the very bottom of the memory that is available for data and works upward. When the compiler is done, the stack assignments are complete and permanent. All remaining memory that is available for data is the heap. Space in the heap is assigned from the top down, by functions such as *malloc* and *calloc*. The heap is often called dynamic memory, since you can call it when you need it and then release it when you are through with it, whereas variables assigned to the stack have permanent memory assignments for the life of the program. It's nice to have some idea of how these memory assignment techniques work, but the compiler and program manipulations of the stack and heap will not directly affect your programming.

The *free* Function

The *free* function is used to release the memory assigned by *malloc* or *calloc* when you are finished using it. If you can release the memory occupied by one of your big arrays, when the operations on this array are done, then the memory will be available for another array that you may need later in the program. This is very important when you are using a lot of memory and spare memory is scarce. The function works like this. First you must have allocated memory by a statement like

```
c = malloc(32000);
```

or

```
d = calloc(15670, 2);
```

To release this memory after you are through using it, you would write

```
free(c);
free(d);
```

The *farmalloc* and *farcalloc* Functions

While *malloc* and *calloc* are limited by the size of the memory model used, *farmalloc* and *farcalloc* can make use of all memory available to the computer. These functions make use of unsigned long arguments rather than the unsigned arguments used by *malloc* and *calloc*. Otherwise, the two pairs of functions are essentially the same for compact, large and huge memory models. For tiny memory models, the *farmalloc* and *farcalloc* functions cannot be used. For small and medium memory models, they can access greater than 64K blocks of memory, which is quite different from *malloc* and *calloc*, which are limited to 64K blocks. The functions are used in the same way as the *malloc* and *calloc* functions. First, suppose we have the code

```
char far * bigarray;
...
...
if ((bigarray = farmalloc(88000)) == NULL)
{
        printf("\nNot enough memory for bigarray...");
        getch();
        exit(0);
}
...
...
value = bigarray[77000];
```

This **will not** work. There is nothing wrong with assigning 88000 bytes for the size of *bigarray*, but in the last statement, you cannot have an array member larger than 64K bytes, regardless of how much space is allocated to the array. You can do this, however

```
movedata(FP_SEG(bigarray) + 4812,
        FP_OFF(bigarray    +    8),     FP_SEG(value),
        FP_OFF(value), 1);
```

We'll go into detail on how to use the *movedata* function later. What we're showing here is that for array members greater than 64K, you must use the *movedata* function rather than the simple assignment statement that we are used to using.

Even if we are using array members less than 64K, we need to be careful in working with arrays that are represented by far pointers. For example, we could not write

```
strcpy (string,bigarray);
```

nor could we write

```
printf("%s", bigarray);
```

since the string copying function *strcpy* and the display function *printf* assume near pointers for all arguments. Therefore we would have to use the *movedata* function for the *strcpy* operation and would have to transfer data from *bigarray* to a location that *printf* would understand before using it.

Finally, whatever you do, don't do this:

```
char far *firstarray, *secondarray;
...
...
if ((firstarray = farmalloc(88000)) == NULL)
{
      printf("\nNot enough memory for"
            " firstarray...");
      getch();
      exit(0);
}
if ((secondarray = farmalloc(29000)) == NULL)
{
      printf("\nNot enough memory for"
            " secondarray...");
      getch();
      exit(0);
}
```

If there is enough memory, *farmalloc* will assign memory areas for each array and there will be no error message. However, the very first statement

```
char far *firstarray, *secondarray;
```

despite what you might think, does not associate the word *far* with both pointers. Only the first pointer is a far pointer; the second one is a near pointer. When the far address is returned to it, part of it gets lost, so that who knows where the pointer is addressing memory. When you begin to load stuff into this second array, your program will go totally bananas. What you need to do to be safe is this

```
char far *firstarray;
char far *secondarray;
```

Then everything will work all right.

The *farfree* Function

The *farfree* function is used to release the memory assigned by *farmalloc* or *farcalloc* when you are finished using it. As with memory areas assigned by *malloc* and *calloc*, if you can release the memory occupied by one of your big arrays, when the operations on this array are done, then the memory will be available for another array that you may need later in the program. This is very important when you are using a lot of memory and spare memory is scarce. The function works like this. First you must have allocated memory by a statement like

```
c = farmalloc(32000);
```

or

```
d = farcalloc(15670, 2);
```

To release this memory after you are through using it, you would write

```
farfree(c);
farfree(d);
```

The *free* and *farfree* functions are not interchangeable. If you assigned memory with the *malloc* or *calloc* functions, you must release it with the *free* function. If you assigned memory with *farmalloc* or *farcalloc* you must release it with the *farfree* function. If you select the wrong function to release a memory block, the program will quit and give you an error message.

The *memchr* Function

The *memchr* function searches through a block of memory and returns a pointer to the first occurrence of a designated character. If the character does not appear, a NULL is returned. The function is used in this way

```
char string[20], * pointer;
...
...
pointer = (char *) memchr(string, 'd',
      strlen(string))
```

At the beginning of the program, we define the character array *string*, which contains the array to be searched, and also the pointer, *pointer*, which will contain the result of the search. Later in the program, we call the function *memchr*. We must type cast the return from this function to match the type of the pointer that is to receive the result by preceding the function name with *(char *)*. The first argument passed to this function is a pointer to the string that is to be tested. The second argument is the character that is being searched for. (You can give the actual character in single quotation marks, or the ASCII number of the character, or the number of the character in hexadecimal preceded by *0x*.) The third argument is the size of the memory block to be searched. In this example, we have used the *strlen* function to set the size to the actual size of the string contained in the string array. (You could just specify the size of the array, if you're sure there isn't any garbage following the end of the actual string that would mess up your result. Here is an example of how you can foul yourself up in this way, however

```
#include <stdio.h>
#include <string.h>
#include <mem.h>
#include <conio.h>

void main(void)
{
       char str[40] = {"This is the longest"
              " string..."}, *pointer;
       strcpy(str,"This is a string");
       printf("\n%s",str);
       pointer = (char *) memchr(str, '.', 40);
       printf("\n%p",pointer);
       pointer = (char *) memchr(str, '.',
       strlen(str));
       printf("\n%p",pointer);
       getch();
}
```

What happens here is that the *str* array is initialized with *This is the longest string...* followed by NULLs to the end of the array. When we use *strcpy* to replace the initial string with a new one, the new string is written to the array and terminated with a NULL, but any part of the old string that follows the NULL is not zeroed; it just remains there. When we run *memchr* for the whole length of the array, it reports back the location of the first period encountered, which is the first period in the garbage remaining in the array from the old string. This isn't what we want. Since the new string doesn't contain a period we want the function to return a NULL. When we run *memchr* again, with the length of the memory block specified as the actual length of the current string, the search stops at the end of the string and we obtain the desired result.)

Here's a sample program fragment that shows how you might want to use the *memchr* function.

```
char filename[13], *pointer;
int length;
...
...
length = strlen(filename);
if (length > 9)
      length = 9;
pointer = (char *) memchr(filename, '.', length);
if (pointer == NULL)
{
      if (length > 8)
            length = 8;
      strncpy(filename,filename,length);
      filename[length+1] = NULL;
      strcat(filename,".EXT");
}
```

In this piece of code, we assume that a DOS file name has been entered into the array *filename* in the part of the program that is not shown. A legal DOS file name may consist of 1 to 8 characters which may optionally be followed by a period and an extension of up to 3 characters. If the file name already includes an extension, we're going to leave it alone. If it doesn't, we're going to add the period and extension *EXT*. We first set *length* to the actual length of the file name or to 9 characters, whichever is smaller. (For the file name to be legitimate, if it is longer than 8 characters, the ninth character must be a period.) We then use *memchr* to check whether a period is present. If it is not, *pointer* will contain a null. If this is the case, we first use *strncpy* to truncate the file name to 8 characters if it is greater than that. (Just in case someone typed in an illegal file name containing more than 8 characters, we're going to make it a legal 8 characters.) Then we use *strcat* to append the period and the extension *EXT* to the end of the file name.

The *memchr* function uses near pointers so the memory location tested must be a near one. Therefore you could not, for example, use it to check display memory for a desired character. Some compilers have a similar function that can be used with far memory; others don't. Make sure your compiler has such a function before you attempt to use it, and don't use it at all if you want your program to be portable to many C compilers.

The *memcmp* and *qsort* Functions

The *memcmp* function compares two blocks of memory. Three arguments are passed to the function. The first is a pointer to the first block of memory, the second is a pointer to the second block of memory and the third is the number of bytes to be compared. The function would be used like this

```
char buffer1[30], buffer2[35];
int result;
...
...
result = memcmp(buffer1, buffer2,25);
if (result > 0)
        printf("Buffer 1 is larger");
if (result < 0)
        printf("Buffer 2 is larger");
if (result == 0)
        printf("Buffer 1 and Buffer 2 are the same");
```

In this example, *memcmp* compares the first 25 characters of the two buffers, and the result is based on these 25 characters only. If the first item is larger than the second, the function returns a number larger than zero. If the two items are equal, the function returns a zero. If the second item is larger than the first, the function returns a number less than zero.

One of the uses that you will find for the *memcmp* function is in sorting lists. The function *qsort* is included with many compilers. It performs a variant of the quicksort sorting technique. As used in C, *qsort* is passed four arguments, the address of the list to be sorted, the number of items in the list, the size of each item, and the name of a comparison function like this

```
qsort(char * list, 35, 14, sort_function);
```

The *qsort* function sorts by sending pairs of items to the sort function. If the sort function returns a positive number or zero, *qsort* leaves the pair as it is. If the sort function returns a negative number, *qsort* interchanges the two numbers. This gives you a lot of flexibility in defining your own sort function. You can set it up so as to sort in ascending or descending order. Figure 17-1 lists a program that gives an example of sorting a list of words and a list of numbers using the *qsort* function.

Figure 17-1. Program to Demonstrate Sorting

```
/*

        SORT = Program to demonstrate qsort function

                By Roger T. Stevens  8-9-92

*/

#include <stdlib.h>
#include <stdio.h>
#include <mem.h>
#include <conio.h>

char list[6][10] = {"once", "twice", "afterward", "soon",
        "forever", "never"};
int numbers[8] = {123,67,45,2,13,76,8,4};

int sort1(const void *a, const void *b);
int sort2(const void *a, const void *b);

void main(void)
{
        int i;

        clrscr();
        for (i=0; i<6; i++)
                printf("\n%s",list[i]);
        getch();
        printf("\n\n");
        qsort(list, 6, sizeof(list[0]), sort1);
        for (i=0; i<6; i++)
                printf("\n%s",list[i]);
        getch();

        printf("\n\n");
        qsort(numbers, 8, sizeof(int), sort2);
        for (i=0; i<8; i++)
                printf("\n%d",numbers[i]);
        getch();
}
```

(continued)

```
/*
    sort1() = Sort function for character strings
*/

int sort1(const void *a, const void *b)
{
      return(memcmp((char *) a, (char *) b, 9));
}

/*
    sort2() = Sort function for integers
*/

int sort2(const void *a, const void *b)
{
      return(memcmp((int *) a, (int *) b, sizeof(int)));
}
```

You should have no trouble with the *qsort* function if you're careful about a couple of things. First any sort function you create (such as *sort1* or *sort2*) must have two arguments, each of the type *const void* *. This is a generic way of specifying a pointer in ANSI C, and is the way in which *qsort* is designed in Borland C++, Version 3.1. If you are using another compiler, *qsort* may be defined so that your sort function requires arguments of a different type. Check your reference manual to make sure you are specifying the right type of arguments for your sort function. (Even Borland has been inconsistent on this so that earlier versions of Turbo C, etc. require a different definition of the sort function.) Second, make sure that the data item length specified in *qsort* is the actual length in bytes of the data items. To make sure, you can specify it using the *sizeof* function as is done in the example program. You might suppose that if you have an array of strings, each of which is 8 bytes long, but you are actually using only 4 bytes for each string, you could specify a length of 4 and everything would be fine, but this is not true; after you sorted in this way, you would be left with a list containing only fragments of the original data. Furthermore, the size must count the NULL terminating character of a string, so that the right number would be 9 for this case.

The program is pretty simple. It defines a list of words and a list of integers. It then prints out the list of words. When you hit a key, the program sorts the list of words and prints out the sorted list. When you hit a key again, the program sorts the list of integer numbers and prints out the sorted list.

As with the *memchr* function, the *memcmp* function uses near pointers so the memory location tested must be a near one. Therefore you could not use it to compare blocks of far memory. Some compilers have a similar function that can be used with far memory; others don't. Make sure your compiler has such a function before you attempt to use it, and don't use it at all if you want your program to be portable to many C compilers.

The *memset* Function

The *memset* function is used to set a block of memory to a designated character. It has three parameters, the address of the memory block, the character to which the memory is to be set, and the number of bytes in the block. It is a good function to use with *malloc* to initialize a block of memory when you want it initialized to something other than 0. (If 0 is your choice, you can do the job in a single operation with *calloc* as previously described.) The function would work in this way

```
char * array;
...
...
array = malloc(12000);
memset(array,'k',12000);
```

This would initialize the block of memory with the character *k* (0x6B).

The *movedata* Function

The *movedata* function is an extremely useful function; it can move a block of memory from one location to another even if the source or destination or both are designated by far pointers. It can be used with memory blocks greater than 64K bytes. In fact, when you have used *farmalloc* or *farcalloc* to allocate a memory block larger than 64K bytes, *movedata* is often the only way that you have to move data to and from this block. The function has five parameters, a pointer to the source segment address, a pointer to the source offset address, a pointer to the destination segment address, a pointer to the destination offset address and the number of bytes in the block to be moved. The most convenient way to get these pointers is through the use of the *FP_SEG* and *FP_OFF* functions which get the segment and offset values of a far pointer. (Although these functions were designed to be used with pointers to far memory, they work just as well for any address, near or far.) A typical *movedata* statement would look like this

```
movedata(FP_SEG(source), FP_OFF(source),
      FP_SEG(destination), FP_OFF(destination,
      4500);
```

Figure 17-2 is a sample program that reads picture data from a disk file to a buffer and then uses *movedata* to transfer it to the display screen. You've seen a similar program in Chapter 11. The version used here is stripped to the very minimum and is very fast. It makes use of four passes through a *for* loop. Each one reads the complete data for one color plane into a buffer. An output to the VGA register sets up the VGA to accept data for the appropriate color plane. The *movedata* statement then transfers the data to the display.

Figure 17-2. Program to Read Data from Disk and Display It

```
/*

    READSCRN = Program to read a graphics information
         disk file and display it on the screen.

              By Roger T. Stevens  8-12-92

*/

#include <stdio.h>
#include <dos.h>
void setmode(int mode);

FILE *fsave;
char buffer[38400];
int i;
```

(continued)

```
main()
{
      setmode(18);
      fsave = fopen("tcheb.raw","rb");
      for (i=0; i<4; i++)
      {
            fread(buffer,1,38400,fsave);
            outport(0x3C4,(0x01<<(8+i)) + 2);
            movedata(FP_SEG(buffer),FP_OFF(buffer),
                  0xA000,0,38400);
      }
      fclose(fsave);
      getch();
}

/*
            setmode() = Sets video mode
*/

void setmode(int mode)
{
      union REGS reg;

      reg.x.ax = mode;
      int86 (0x10,&reg,&reg);
}
```

Offsetting Using *movedata*

We've pointed out that if you have a character array of, for example, 75,000 bytes, you can't use a statement such as

```
a = buffer[73200];
```

to access a high member of it, since the highest number that can be used to enumerate a member of an array is 64K. Thus we must use the *movedata* statement. We might think that we could do this by

```
movedata(FP_SEG(buffer), FP_OFF(buffer) + 73200,
      FP_SEG(&a) FP_OFF(&a), 1);
```

but this is not possible because the offset value cannot be greater than 64K. Furthermore, we don't know what value will be returned by *FP_OFF*, since it depends on the memory location where the array begins, so that even if the offset we wanted to add was a lot smaller than 73,200, we might end up with an offset value greater than 64K. The safest way to do this offsetting is this:

```
movedata(FP_SEG(array + 73200L), FP_OFF(array +
        73200L), FP_SEG(&b),FP_OFF(&b),1);
```

This technique allows the *FP_SEG* and *FP_OFF* functions to go directly to the offset address in the array and return the segment and offset values broken up in a way that is acceptable to the computer system.

Creating Plasma Mountains

The *plasmamt* program has three separate phases of operation. The first generates a plasma display. This is essentially the same as the *plasma* program given in Chapter 12. The entire display is then saved to a buffer. In the second phase, the plasma display is considered to be a contour map, with the color corresponding to height. The resulting three-dimensional information is projected to a three-dimensional representation, from a specified viewing angle, on the two-dimensional display screen, using the same color representation. Finally, in the third phase, this three-dimensional projection is colored according to the light falling upon it from a single light source at a specified location. This is the most complex program in this book. Looking at it, you may have a tendency to get discouraged and say, "How can I ever figure out how to write my own program if it needs this much complexity?" Don't get discouraged, however. First, you'll note that this program builds on some of the other programs that we have used previously, so that much of the work was already done and only had to be modified to work in the new application. Second, when you write your own complex program, you will be completely knowledgeable on the math and procedures involved, because it is your field of expertise. The only thing new to you will be converting it to a C program. If you fully understand everything else, you won't find it too difficult to write the software, using the techniques described in this book and tackling the overall problem one step at a time.

The program for generating plasma mountains is listed in Figure 17-3. Plate 14 shows the resulting plasma mountain after the realistic coloring algorithms are applied. The program begins by setting the display mode to the 320 × 200 pixel × 16 color graphics mode. It then calls *initpalette* to initialize the 256 VGA color registers for the plasma display. Next, the program uses *farcalloc* to set up *buffer*, an array of 64,000 characters; *col_buffer1* and *colbuffer2*, arrays of 33,000 integers each; *col_hist*, an array of 32,768 characters; *hues*, an array of 8192 integers; and *frequency*, an array of 8192 characters. If memory is not available for any of these arrays, the program terminates with an error message.

The program next proceeds as did the *plasma* program of Chapter 11, first plotting four randomly colored points at the corners of the screen and then recursively calling the function *subdivide* to plot points closer and closer together until the entire screen is filled.

When the plasma display is complete, the program uses the *movedata* statement to move the entire display memory to the array *buffer*. The program then pauses and waits for a keystroke.

Once a keystroke occurs, the program proceeds to the next phase. It begins by converting the light source and viewing angles into radians and computing the sine and cosine of the viewing angles. The angles are initialized at the beginning of the program; if you want to try changing them, you can go to that point and insert new angles in degrees. The program then computes vectors representing the light and viewing directions. We won't go into all the details here of how to project a light beam on the scene or how to determine the way the viewing position affects the projection of the three-dimensional scene onto the two-dimensional display screen. If you want to pursue this, you can refer to *The C Graphics Handbook*. Here, we are mostly interested in seeing how the program uses memory management. The program next sets the values for the *ambient, diffuse, specular, background*, and *alt_background* colors. The first three represent the colors that make up the mountains. The *background* color is the color of the sky. The *alt_background* color is the color of the lake at the base of the mountains. The program makes the integer color values for the *background* and *alt_background* colors and sets their frequency of occurrence to the maximum (255) so that they will be sure to be included in the palette of available colors.

Next, the program begins the process of painting the three-dimensional projection onto the screen. First, each pixel in this column from row 0 to row 164 in *color_buf1* and *color_buf2* (where the final color information is stored) is set to the background or sky color so that it will appear as sky if it is not set to a mountain color. Next, the program enters a *for* loop that iterates once for each data column on the screen. This loop begins by taking the top row height value for the plasma at the current column and computing where it will appear on the screen for the projection from three to two dimensions. This value is placed in *horizon*. Next an inner *for* loop is entered, which scans down the screen from top to bottom. At each iteration the program gets the plasma height for this column and row and determines the projected location. If this location is greater than the *horizon* value, the point is plotted to the screen. The function *intensity* is also called. This function determines a normal vector to the surface at this point and uses it and the light vector to determine the mountain color for this pixel. Next, if there is a gap between this point and the previously plotted point on this column, the program interpolates colors and plots points between this and the previous point until the screen is filled between them. As each point is plotted, a color number is computed for it. This is determined by reducing the possible number of shades of red, green, and blue from 64 each to 32 each, and combining them into a single integer. This amounts to 5 bits for each primary color. The most significant bit of the integer is not used; the five next most significant bits are the blue color, followed by five bits for the green color and the five least significant bits are the red color. This color number goes to the proper location in one of two buffers, *col_buffer1* or *col_buffer2*. These together contain the 64,000 integers needed to save all the points on the display. In order to be able to access each point directly, by the array member number, we've used two smaller buffers rather than one large one. In addition, the member of the *color_hist* array corresponding to this color number is incremented to keep account of the number of times each color occurs. Finally the rows from row 165 to row 199 are changed to the alternate background or lake color. In addition to providing a lake at the base of the mountains, this covers up any irregular gaps that were left at the bottom of the mountains. At the end of this second phase of the program, the three-dimensional projection is displayed in the original colors and all of the color information for the final illuminated mountain display is stored in the *color_buf1* and *color_buf2* arrays.

We're now through with the *buffer* array, so we call the function *farfree* to release this memory back to the computer. The next thing that occurs is that the *memset* function is used to zero out the palette array, so that new data can be stored in it. The program now scans through the *color_hist* array and, for each color that occurs at least once, places the color number in the next adjacent member of the *hues* array and the number of occurrences (up to 255) in the corresponding member of the *frequency* array. (These arrays can hold a maximum of 8192 colors. Actually, for the pictures painted by this program, fewer than 100 colors are required.) Next, a special version of the quicksort algorithm is called (the function *sort*), which sorts the members of the *frequency* array in descending order and at the same time keeps the color number for each color in the matching member of the *hues* array. The first 256 colors are now put into the *palette* array and are then used to set color registers 1 to 255 of the VGA. (Color register 0 is set to black.) The *setmode* function is also called. Its main purpose here, since the mode is not changed, is to blank the screen. The *color_hist* array is zeroed with the *memset* function and then the members of this array corresponding to colors that were set in the 256 color registers are set to the proper register numbers. Now we can take an integer color number and by selecting the contents of the corresponding member of the *color_hist* array we obtain the number of the color register where this color is stored.

In the final phase, the program enters two nested *for* loops. The outermost iterates once for each column of the display. The inner loop iterates for each row of the display. Together, they cover every pixel on the screen. For the first 160 columns of the screen, the program uses the *color_hist* array to get the proper color register number for the color number stored at the current pixel location in *color_buf1*. This color is displayed at the pixel location on the screen. For the remaining columns, the same information is obtained from *color_buf2* and displayed. The resulting display remains on the screen until a key is struck, at which point the program terminates.

This program should give you a good idea of how the *memset, movedata, farcalloc*, and *farfree* functions are used in a typical application. If you start thinking about how you would substitute other functions to perform these memory management operations, you'll find that you would have to use a lot of disk accesses to substitute for some of them, while others would require time-consuming *for* loops, or more complicated constructions. A little examination of the memory management operations in this program should give you a good appreciation for how useful these functions can be in simplifying and speeding up your programs.

Figure 17-3. Program to Generate Plasma Mountains

```
/*
 ┌──────────────────────────────────────────────────────────────┐
 │                                                              │
 │  plasmaMT = Creates a plasma mountain display an a VGA       │
 │                                                              │
 │             By Roger T. Stevens   8-2-92                     │
 │                                                              │
 └──────────────────────────────────────────────────────────────┘
*/

#include <dos.h>
#include <stdio.h>
#include <stdlib.h>
#include <time.h>
#include <conio.h>
#include <alloc.h>
#include <math.h>

union REGS reg;
struct SREGS inreg;

typedef struct
{
   float x, y, z;
} Vector;

typedef struct
{
   int x, y, z;
} VecInt;

void initPalette(void);
VecInt intensity(void);
void plot(int x, int y, int color);
int readPixel(int x, int y);
void setmode(int mode);
Vector setVec(float x, float y, float z);
void setVGApalette(unsigned char *buffer, int start, int
   regs);
void setVGAreg(int reg_no, int red, int green, int blue);
void sort(unsigned int start, unsigned int end);
void subdivide(int x1, int y1, int x2, int y2);

char far *address;
unsigned char far *color_hist;
unsigned char far *buffer;
unsigned char far *frequency;
int far *col_buffer1;
int far *col_buffer2;
unsigned int far *hues;
float     ViewPhi = 55,
```

(continued)

```
         LightPhi = 75,
         LightTheta = 300;
float SinViewPhi, CosViewPhi,h;
unsigned char d2, height, old_height, new_height,
   palette[512][3];
int b=2, backcol, altbackcol, i, last_color, x1, xres=319,
   yres=199, p, p1, horizon;
long int color_no;
unsigned int j, temp_d;
Vector view, light, ambient, diffuse, specular,
   background, alt_background;
VecInt color_vec, Pix, newPix, oldPix;

void main()
{
   setmode(0x13);
   initPalette();
   if ((buffer = farcalloc(64000,1)) == NULL)
   {
      printf("\nNot enough memory for buffer...\n");
      exit(0);
   }
   if ((col_buffer1 = farcalloc(33000,2)) == NULL)
   {
      printf("\nNot enough memory for first color"
         " buffer...");
      getch();
      exit(0);
   }
   if ((col_buffer2 = farcalloc(33000,2)) == NULL)
   {
      printf("\nNot enough memory for second color"
         " buffer...");
      getch();
      exit(0);
   }
   if ((color_hist = farcalloc(32768,1)) == NULL)
   {
      printf("\nNot enough memory for color"
         " histogram...");
      getch();
      exit(0);
   }
   if ((hues = farcalloc(8192,2)) == NULL)
   {
      printf("\nNot enough memory for hues"
         " array...");
      getch();
      exit(0);
   }
   if ((frequency = farcalloc(8192,1)) == NULL)
   {
      printf("\nNot enough memory for frequency"
```

(continued)

```
            " array...");
        getch();
        exit(0);
    }
    randomize();
    plot(0,0,random(255) + 1);
    plot(xres,0,random(255) + 1);
    plot(xres,yres,random(255) + 1);
    plot(0,yres,random(255) + 1);
    subdivide(0,0,xres,yres);
    movedata(0xA000,0,FP_SEG(buffer),
        FP_OFF(buffer),64000L);
    getch();
    setmode(0x13);
    initPalette();
    ViewPhi *= 0.0174533;
    LightPhi *= 0.0174533;
    LightTheta *= 0.0174533;
    SinViewPhi = sin(ViewPhi);
    CosViewPhi = cos(ViewPhi);
    view = setVec(0, -CosViewPhi, SinViewPhi);
    light = setVec(sin(LightTheta) *
        cos(LightPhi),sin(LightTheta) *
        sin(LightPhi), cos(LightTheta));
    ambient = setVec(0.5, 0.3, 0.2);
    diffuse = setVec(1.0, 0.8, 0.4);
    specular = setVec(1.0, 0.8, 0.4);
    background = setVec(0.6, 0.6, 1.0);
    alt_background = setVec(1.0, 0.8, 0.4);
    alt_background = setVec(0, 0, 0.5);
    backcol = (((int)(background.x * 63) & 0x3e) >> 1) |
        (((int)(background.y * 63) & 0x3e) << 4) |
        (((int)(background.z * 63) & 0x3e) << 9);
    altbackcol = (((int)(alt_background.x * 63)& 0x3e) >>
        1) | (((int)(alt_background.y * 63) & 0x3e) << 4) |
        (((int)(alt_background.z * 63) & 0x3e) << 9);
    color_hist[backcol] = 255;
    color_hist[altbackcol] = 255;
    for (i=0; i<=xres; i++)
    {
        horizon = (float)buffer[i] * CosViewPhi;
        for (j=1; j<165; j++)
        {
            if (i<160)
                col_buffer1[i*200 + j] = backcol;
            else
                col_buffer2[(i-160)*200 + j] =
                    backcol;
        }
        for (j=1; j<=yres; j++)
        {
            if (i>0)
                old_height = buffer[j*320 + i - 1];
```

(continiued)

```
        height = buffer[j*320 + i];
        new_height = buffer[(j+1)*320 + i];
        p1 = (float)(j) * SinViewPhi +
           (float)(height) * CosViewPhi;
        if (p1 >= horizon)
        {
           oldPix = intensity();
           if ((p1 <= yres) && (p1 >= 0))
           {
              plot(i, p1, height);
              color_no = ((oldPix.x & 0x3e) >> 1) |
                 ((oldPix.y & 0x3e) << 4) | ((oldPix.z
                 & 0x3e) << 9);
              if (i<160)
                 col_buffer1[i*200 + p1] = color_no;
              else
                 col_buffer2[(i-160) * 200 + p1] =
                    color_no;
              if (color_hist[color_no] < 255)
                 color_hist[color_no]++;
                 p = p1 - 1;
              newPix = intensity();
              while (p >= horizon)
              {
                 h = (float)(p - horizon) / (float)(p1
                    - horizon);
                 Pix.x = ((oldPix.x * h) + (newPix.x *
                    (1 - h)));
                 Pix.y = ((oldPix.y * h) + (newPix.y *
                    (1 - h)));
                 Pix.z = ((oldPix.z * h) + (newPix.z *
                    (1 - h)));
                 if ((p <= yres) && (p >= 0))
                 {
                    plot(i, p, height);
                    color_no = ((Pix.x & 0x3e) >> 1) |
                       ((Pix.y & 0x3e) << 4) |
                       ((Pix.z & 0x3e) << 9);
                    if (i<160)
                       col_buffer1[i*200 + p] =
                          color_no;
                    else
                       col_buffer2[(i-160)*200 + p] =
                          color_no;
                    if (color_hist[color_no] < 255)
                       color_hist[color_no]++;
                 }
                 p--;
              }
           }
           horizon = p1;
        }
    }
```

(continued)

```
        for (j=165; j<=yres; j++)
        {
            if (i<160)
                col_buffer1[i*200 + j] = altbackcol;
            else
                col_buffer2[(i-160)*200 + j] = altbackcol;
        }
    }
    getch();
    farfree(buffer);
    memset(palette,0x00,768);
    last_color = -1;
    for (j=0; j<32768; j++)
    {
        if (color_hist[j] > 0)
        {
            last_color++;
            hues[last_color] = j;
            frequency[last_color] = color_hist[j];
        }
    }
    sort(0,last_color);
    for (j=1; j<256; j++)
    {
        palette[j][0] = (hues[j] << 1) & 0x3E;
        palette[j][1] = (hues[j] >> 4) & 0x3E;
        palette[j][2] = (hues[j] >> 9) & 0x3E;;
    }
    setmode(0x13);
    setVGApalette(&palette[0][0],1,255);
    setVGAreg(0,0,0,0);
    memset(color_hist, 0x00, 32768);
    for (j=0; j<=last_color; j++)
        color_hist[hues[j]] = j + 1;
    for (x1=0; x1<=xres; x1++)
    {
        for (i=0; i<=yres; i++)
        {
            if (x1<160)
                color_no = col_buffer1[x1*200 + i];
            else
                color_no = col_buffer2[(x1-160)*200 + i];
            plot(x1,i,color_hist[color_no]);
        }
    }
    getch();
    setmode(3);
}
```

(continued)

```
/*
    initpalette() = Sets the colors of the VGA palette
*/

void initPalette(void)
{
   int max_color = 63;
   int index;

   for (index=0; index<85; index++)
   {
      palette[index][0] = 0;
      palette[index][1] = (index*max_color) / 85;
      palette[index][2] = ((86 - index)*max_color) / 85;
      palette[index+255][0] = 0;
      palette[index+255][1] = (index*max_color) / 85;
      palette[index+255][2] = ((86 - index)*max_color) /
         85;
      palette[index+85][0] = (index*max_color) / 85;
      palette[index+85][1] = ((86 - index)*max_color) /
         85;
      palette[index+85][2] = 0;
      palette[index+339][0] = (index*max_color) / 85;
      palette[index+339][1] = ((86 - index)*max_color) /
         85;
      palette[index+339][2] = 0;
      palette[index+170][0] = ((86 - index)*max_color) /
         85;
      palette[index+170][1] = 0;
      palette[index+170][2] = (index*max_color) / 85;
      palette[index+424][0] = ((86 - index)*max_color) /
         85;
      palette[index+424][1] = 0;
      palette[index+424][2] = (index*max_color) / 85;
   }
   setVGAreg(0,0,0,0);
   setVGApalette(palette[0],1,255);
}

/*
  intensity() = Sets value of a color vector for a pixel
*/

VecInt intensity(void)
{
   Vector normal, reflected;
```

(continued)

```
    float CosTheta, CosAlpha, temp;

    normal.x =  (float)(height - new_height) * 3.072;
    normal.y = 2.7136 * (float)(old_height - height);
    normal.z = 8.336;
    temp = sqrt(normal.x * normal.x + normal.y * normal.y +
       normal.z * normal.z);
    if (temp != 0)
    {
       normal.x /= temp;
       normal.y /= temp;
       normal.z /= temp;
    }
    CosTheta = normal.x * light.x + normal.y * light.y +
       normal.z * light.z;
    if (CosTheta < 0)
    {
       color_vec.x = ambient.x * 63;
       color_vec.y = ambient.y * 63;
       color_vec.z = ambient.z * 63;
    }
    else
    {
       reflected.x = (normal.x - light.x) * (2.0*CosTheta);
       reflected.y = (normal.y - light.y) * (2.0*CosTheta);
       reflected.z = (normal.z - light.z) * (2.0*CosTheta);
       CosAlpha = view.x * reflected.x + view.y *
          reflected.y + view.z * reflected.z;
       color_vec.x = ((specular.x * pow(CosAlpha, b) +
          ambient.x * diffuse.x + (diffuse.x * CosTheta))
          * 32);
       color_vec.y = (specular.y * pow(CosAlpha, b) +
          ambient.y * diffuse.y + (diffuse.y * CosTheta))
          * 32;
       color_vec.z = (specular.z * pow(CosAlpha, b) +
          ambient.z * diffuse.z + (diffuse.z * CosTheta))
          * 32;
    }
    return(color_vec);
}

/*

    plot() = Function to plot a point to VGA 256 color
                          screen

*/

void plot(int x, int y, int color)
{
    unsigned int offset;
    char far *address;
```

(continued)

```
    offset = 320 * y + x;
    address = (char far *)(0xA0000000L + offset);
    *address = color;
}

/*
┌──────────────────────────────────────────────────────┐
│                                                      │
│        readPixel() = Read a pixel from the screen    │
│                                                      │
└──────────────────────────────────────────────────────┘
*/

int readPixel(int x, int y)
{
    reg.h.ah = 0x0D;
    reg.x.cx = x;
    reg.x.dx = y;
    int86 (0x10,&reg,&reg);
    return (reg.h.al);
}

/*
┌──────────────────────────────────────────────────────┐
│                                                      │
│               setmode() = Sets video mode            │
│                                                      │
└──────────────────────────────────────────────────────┘
*/

void setmode(int mode)
{
    reg.x.ax = mode;
    int86 (0x10,&reg,&reg);
}

/*
┌──────────────────────────────────────────────────────┐
│                                                      │
│        setVec() = Sets the value of a vector         │
│                                                      │
└──────────────────────────────────────────────────────┘
*/

Vector setVec(float x, float y, float z)
{
    Vector temp;

    temp.x = x;
    temp.y = y;
    temp.z = z;
    return(temp);
}
```

(continued)

```
/*
    setVGApalette() = Sets a specified set of VGA color
                         registers
*/

void setVGApalette(unsigned char *buffer,int start, int
   regs)
{
   reg.x.ax = 0x1012;
   segread(&inreg);
   inreg.es = inreg.ds;
   reg.x.bx = start;
   reg.x.cx = regs;
   reg.x.dx = (int)&buffer[0];
   int86x(0x10,&reg,&reg,&inreg);
}

/*
    setVGAreg() = Sets an individual VGA color register
*/

void setVGAreg(int reg_no, int red, int green, int blue)
{
   reg.x.ax = 0x1010;
   reg.x.bx = reg_no;
   reg.h.ch = red;
   reg.h.cl = green;
   reg.h.dh = blue;
   int86(0x10,&reg,&reg);
}

/*
    sort() = Quicksort to sort colors by frequency
*/

void sort(unsigned int start, unsigned int end)
{
   unsigned int pivot,temp2;
   unsigned char temp;

   if (start < (end - 1))
   {
      i = start;
```

(continued)

```
        j = end;
        pivot = (frequency[i] + frequency[j] +
           frequency[(i+j)/2])/3;
        do
        {
           while (frequency[i] > pivot)
              i++;
           while (frequency[j] < pivot)
              j--;
           if (i < j)
           {
              temp = frequency[i];
              frequency[i] = frequency[j];
              frequency[j] = temp;
              temp2 = hues[i];
              hues[i++] = hues[j];
              hues[j--] = temp2;
           }
        }  while (i < j);
        if (j < end)
        {
           sort(start,j);
           sort(j+1,end);
        }
     }
     if (frequency[end] > frequency[start])
     {
        temp = frequency[start];
        frequency[start] = frequency[end];
        frequency[end] = temp;
        temp2 = hues[start];
        hues[start] = hues[end];
        hues[end] = temp2;
     }
}

/*

   subdivide() = Divides up a display section and fills
                         with color

*/

void subdivide(int x1, int y1, int x2, int y2)
{
   int x, y, dist, color,rp11, rp12, rp21, rp22;

   x = (x1 + x2) >> 1;
   y = (y1 + y2) >> 1;
   if (x == x1)
      return;
   rp11 = readPixel(x1,y1);
```

(continued)

```
    rp12 = readPixel(x1,y2);
    rp21 = readPixel(x2,y1);
    rp22 = readPixel(x2,y2);
    dist = x2 - x1;
    color = random(dist<<1) - dist;
    color += (rp11 + rp21 + 1) >> 1;
    color = (color < 1) ? 1: (color > 255) ? 255: color;
    if (readPixel(x,y1) == 0)
       plot(x,y1,color);
    color = random(dist<<1) - dist;
    color += (rp12 + rp22 + 1) >> 1;
    color = (color < 1) ? 1: (color > 255) ? 255: color;
    if (readPixel(x,y2) == 0)
       plot(x,y2,color);
    dist = y2 - y1;
    color = random(dist<<1) - dist;
    color += (rp21 + rp22 + 1) >> 1;
    color = (color < 1) ? 1: (color > 255) ? 255: color;
    if (readPixel(x2,y) == 0)
       plot(x2,y,color);
    color = random(dist<<1) - dist;
    color += (rp11 + rp12 + 1) >> 1;
    color = (color < 1) ? 1: (color > 255) ? 255: color;
    if (readPixel(x1,y) == 0)
       plot(x1,y,color);
    color = (rp11 + rp21 + rp22 + rp12 + 2) >> 2;
    plot(x,y,color);
    subdivide(x1,y1,x,y);
    subdivide(x,y1,x2,y);
    subdivide(x,y,x2,y2);
    subdivide(x1,y,x,y2);
}
```

18

Time and Date Functions

The first personal computers built by IBM did not have any way to store time and date information. Every time that you booted up the computer, you had to enter the current date and time. After you did this, the system clock would keep track of the current date and time. You were in good shape until you turned off the power; then the date and time information would go away and have to be inserted again the next time you powered up. This turned out to be very unsatisfactory, so plug-in boards almost immediately began to become available that contained a battery-powered clock which preserved the date and time information while the computer power was off. Such boards caught on quickly and the battery powered clock became a part of the motherboard configuration beginning with the PC AT models. Every modern PC clone that uses a 286 or higher processor has the built-in battery clock, so that current time and date information is always available. It's stored and retrieved in rather odd ways, however. In fact, the basic timing data from the PC is in terms of the number of elapsed seconds since 00:00:00 GMT, January 1, 1970. There are a number of C functions that build upon this basic time information to provide you with different kinds of time and date information. You may not need this too often, but when you do, you'll need to know how to make use of these functions. Several uses of the time and date information are:

(1) You might want to bring up the current time and date automatically to store in a letter or display or to use to tag a file that contains time-sensitive information.

(2) You might want to measure the elapsed time for some part of your program's operation.

(3) You might want to include a time out feature that terminates some particular program action if a response is not received within a specified time.

All of these require the use of the C time and date functions.

The *time* Function

The *time* function is the basic function that returns the number of elapsed seconds from 00:00:00 Greenwich Mean Time (GMT) January 1, 1970. This function, like many of the time associated functions, returns its value to a special data type, *time_t*. If you take a look at the C header files, you will probably find that the data type *time_t* is actually a type *long* (integer). The *time* function can actually be used in either of two ways. This program shows both:

```
#include <stdio.h>
#include <time.h>
#include <conio.h>

void main(void)
{
      time_t time_in_sec;
      time_in_sec = time(NULL);
      printf("\nNo. of sec. since Jan. 1, 1970:"
            " %ld",time_in_sec);
      time_in_sec = 0;
      time(&time_in_sec);
      printf("\nNo. of sec. since Jan. 1, 1970:"
            " %ld",time_in_sec);
      getch();
}
```

In either case, you need an argument of type *time_t*, which in the example is called *time_in_sec*. The first method of retrieving the seconds is to set the argument *time_in_sec* equal to the return from the function *time* and pass a NULL argument to *time*. This results in the time being stored in *time_in_sec*. The other method is to call the *time* function with the address of *time_in_seconds* as the argument. Again, the time is stored in *time_in_sec*. If you run the program, you'll see that both results are the same.

Setting the Time Zone

If you've gone through the process of setting the current local date and time into your PC, you may be wondering how the *time* function just described can determine from the local time and date that you typed in what the Greenwich Mean Time is. After all, your computer may be located anywhere in the world, and just where it is determines the difference between local time and Greenwich Mean Time. The answer is that there is a string called *TZ* in the system environment that contains this important information. This string contains a three character string representing the abbreviation for the current time zone, followed by an optional plus or minus sign, followed by 1 or 2 digits that are the number of hours difference between GMT and your local time, followed by a three character string that is the abbreviation for local daylight savings time. There are two ways that you can set the value of this environmental string. First, you can place a statement like the following in your *autoexec.bat* file, so that it is executed each time you boot up your computer

```
SET TZ="PST8PDT"
```

Alternately, you can use the C *putenv* function to make the change from your program like this

```
putenv("TZ=PST8PDT");
```

You then have to make sure that the function *tzset* is executed as follows to put this information into effect in your C program:

```
tzset();
```

So far, so good. You have now set values for the global variables *daylight, timezone* and *tzname* which are used internally by various of the date and time functions including the *time* function previously described. There are a few things that you need to know about this whole procedure, however. First, the *TZ* string will accept any three-character string as the first string, which gives the abbreviation for the timezone. In the examples, we used *PST*, which you recognize as the abbreviation for Pacific Standard Time. However, we could have entered the string *XXX* and it would have been just fine with the computer. Second, the number following the string is the most important part of TZ. It specifies the hours difference between local time and GMT. You need to be sure to have it right, since this is what functions use to determine the time offset, completely disregarding the first three-character string. Third, although in the example we have used *PDT*, which we understand to be the abbreviation for Pacific Daylight Time, the computer will accept any three-character string for the second three-character string and will accordingly make the assumption that the local time is daylight savings time. If you want local time to be in standard time, you must leave out the second three-character string altogether, ending the *TZ* string after the number of hours offset. Fourth, after you have used *tzset();*, the global variable *tzname[0]* is a pointer to the three-character string that you have entered for the time zone abbreviation and the global variable *tzname[1]* is a pointer to the three-character string that you have entered for the daylight savings abbreviation. If *tzname[1]* points to a NULL string, then standard time has been specified rather than daylight savings time. Finally, if you do not set the *TZ* environment string, it will take on some default value that depends on what BIOS, what version of DOS, and what C compiler you are using. This will usually be *TZ=EST5EDT* but you can't be sure. The best thing you can do is write a short program that uses *getenv* to read *TZ* and then *printf* to display it. Once you find out what the default value is for your system, record it somewhere so you'll always know what it is.

The *ftime* Function

The *time* function simply gives you a raw number of seconds. The *ftime* function is a little more sophisticated, but not much. It requires an argument of type *timeb*. This argument is a structure that looks like

```
struct timeb
{
        long time;
        short millitm;
        short timezone;
        short dstflag;
};
```

You need to define a variable of this type:

```
struct timeb t;
```

Before using *ftime* you need to set up the environment string variable

TZ by using the *putenv* function in the same way that was shown above, for example

```
putenv("TZ=PST8PDT");
```

Now you are ready to use the *ftime* function with a call like

```
ftime(&t);
```

Note that you do not have to use *tzset* to set the time zone parameters with the information in *TZ*, since *ftime* automatically calls *tzset*. Also observe that the argument passed to *ftime* is the address of the structure *t*. After calling *ftime*, you have this data in the structure *t:*

> *t.time* contains the time in seconds from 00:00:00 Greenwich Mean Time on January 1, 1970, to present.
>
> *t.millitim* contains the fractional part of a second that is part of this overall measured time.
>
> *t.timezone* contains the difference in minutes between Greenwich Mean Time and local time. This is obtained from the global variable *timezone,* which is set when *ftime* internally calls *tzset*.
>
> *destflg* contains a 0 if the local time is standard time and a 1 if the local time is daylight savings time.

You can now print out the contents of any member of the structure or use it in any way that you need to in a program. You probably won't find too much use for this function, however, because there are others that will do most time and date jobs more effectively.

The *ctime* Function

If you are willing to accept a predetermined format for displaying your time and date information, you can use the *ctime* function. The *ctime* function returns the day, date, and time information on the present local time as a 26-character ASCII string. The form in which the information is stored is

```
Fri Aug 21 10:24:51 1992\n\0
```

You can then use the *printf* statement to display this string. For everything to work correctly, you need to have previously used *tzset* to set the time difference in seconds between Greenwich Mean Time and local time into the global variable *timezone* and the proper value into the global variable *daylight* to indicate whether or not daylight savings time is in effect. You can then run a short program like this

```
#include <stdio.h>
#include <time.h>

void main(void)
{
      time_t t;

      t = time(NULL);
      printf(\n%s", ctime(&t);
}
```

You first set up a variable *t*, which is of type *time_t*. Then you use the *time* function to get the raw seconds data into *t*. You can then call *ctime* to produce the formatted ASCII string for display. In the example, this is done within the *printf* statement. Please observe that the argument passed to *ctime* is the address of *t*.

The *asctime* Function

The *asctime* function generates a string of 26 characters formatted exactly the same as the string produced by *ctime*. The difference is that instead of generating the time and date information from the raw seconds data obtained by the *time* function, the *asctime* function gets the data from a structure of the type *tm*. This structure is defined by

```
struct tm
{
     int tm_secs; /* seconds */
     int tm_min; /* minutes */
     int tm_hour; /* hour */
     int tm_mday; /* day of the month */
     int tm_mon; /* month */
     int tm_year; /* year (excluding century */
     int tm_wday; /* day of the week */
     int tm_yday; /* day of year (not shown by asctime) */
     int tm_isdst /* is daylight savings time (not shown
          by asctime) */
};
```

With this function, it is your job as programmer to define an argument of this structure type, *time_data*, for example) and fill all of the members of it with the proper values. You can then print out the 26 character formatted string by

```
printf("\n%s",asctime(&time_data);
```

The *gmtime* Function

We just got through telling you that, as programmer, you were responsible for filling the structure of type *tm* used to supply data to the *asctime* function. The function *gmtime* will perform this job for you, if what you want in the structure is Greenwich Mean Time. It takes the raw seconds data returned by the *time* function and breaks it up to supply all of the numbers that are needed for the *tm* structure. Assuming that you have already used *putenv* and *tzset* to properly set up the global environment argument *TZ*, here's a program that will load the *tm* type structure and then use *asctime* to print out time information in GMT.

```
#include <stdio.h>
#include <stdlib.h>
#include <time.h>
#include <dos.h>

void main(void)
{
     time_t seconds;
     struct tm *gmt;

     seconds = time(NULL);
     gmt = gmtime(&seconds);
     printf("\nGMT is: %s", asctime(gmt));
}
```

The *localtime* Function

If you want to fill the structure of type *tm* used to supply data to the *asctime* function with local time data, you can use the function *localtime*. It takes the raw seconds data returned by the *time* function and breaks it up to supply all of the numbers that are needed for the *tm* structure. Assuming that you have already used *putenv* and *tzset* to set up the global environment argument *TZ* properly, here's a program that will load the *tm* type structure and then use *asctime* to print out time information in local time:

```
#include <stdio.h>
#include <stdlib.h>
#include <time.h>
#include <dos.h>

void main(void)
{
     time_t seconds;
     struct tm *local;

     seconds = time(NULL);
     local = localtime(&seconds);
     printf("\nLocal time is: %s", asctime(local));
}
```

Setting System Time from C

The *mktime* function performs two functions. It is passed one argument, the address of a structure of type *tm*. We already described *tm* in connection with the *gmtime* function and pointed out there that it was your responsibility as programmer to insert proper values into it. At that time, we didn't emphasize that you needed to insert legitimate values; if you told *tm* that it was the 30th of February or the time was 25 hours and 72 minutes, for example, you would get a very strange display. The first function of *mktime* is to take a *tm* structure that you have initialized with the values for year, month, day, hour, minute, second, and standard or daylight savings time, and adjust these fields, if they are outside of the legal range of values, so that they are all correct. Your February 30th, for example, would be adjusted to March 2 for an ordinary year or March 1 for a leap year. Your 25:72 time would be adjusted to 2:12 in the morning of the next day. When *mktime* finishes doing this, it computes the proper values to be inserted in the fields for day of the week and day of the year. The second function of *mktime* is to return the elapsed time in seconds from 00:00:00 Greenwich Mean Time on January 1, 1970, for the time and date that have been entered in this structure. This is returned in a data item of type *time_t*. This is the same information that you would get from the system clock by the *time* function. If *mktime* is unable to adjust the *tm* structure to represent the time and date that you have inserted, it returns a -1.

The *stime* function is used to reset the system date and time to a new value. It is passed an argument of type *time_t*, which represents the number of seconds that have elapsed from 00:00:00 on January 1, 1970, Greenwich Mean Time. Ordinarily, you'll want to put in the current time and date. Your program to reset the time would need to put the time you entered into a *tm* structure, convert it from local to GMT, and call the function *mktime* to return the raw seconds number. This would then be passed to *stime* to set the system clock.

The *srfftime* Function

If you aren't content with the standard date and time display format used with *asctime*, you can use the string function *strftime* to create an ASCII string that contains the time data from a structure of type *tm* in just about any way that you want it. This function and its formatting capabilities were described in detail in the previous chapter.

The *difftime* Function

The *difftime* function accepts two arguments of type *time_t*, which are times in seconds. It returns a number of type *double*, which is the difference in seconds between the two arguments. You may wonder a little bit about this function. First, you could easily write a function that would subtract one time from another. Second, since the time difference is always an integral number of seconds, why is a double floating point number used to express this integer? The function does offer one major advantage, however. It doesn't matter what order you use to pass the two times to the function; it always comes up with the absolute time difference between the two.

The Phoenix Fractal

The phoenix fractal is an interesting fractal that continually interchanges real and imaginary parts of the iterated equation and makes use of two previous values. It was discovered by Shigehiro Ushiki. The iterated equations are

$$x_n = x_{n-1}^2 - y_{n-1}^2 + p + qy_{n-2}$$

(Equation 18-1)

$$y_n = 2x_{n-1}y_{n-1} + q\,y_{n-2}$$

(Equation 18-2)

The phoenix curve is perfectly symmetrical about the y axis. The program is used as a demonstration of time functions. We draw the first half of the phoenix curve using floating point operations. Then we switch to an equivalent set of expressions that use only integer operations to do the second half. We measure the time required to produce each half of the display and after the display is done and you hit a key, the time information is shown. If you have a math coprocessor, the floating point half probably takes less time. If you don't have a math coprocessor, you should find that the integer half is much faster. You'll also be able to note slight differences in the structure of the two halves of the display, which result from differences in rounding off techniques. The resulting phoenix display is shown in Plate 15 and the program is listed in Figure 18-1.

The program begins by setting the increments by which x and y are to be increased for each pixel. It then calls the *time* function to get the current time into *first*. The program then enters a pair of nested *for* loops that together include one iteration for every pixel on the left half of the display screen. Within the inner loop, the program initializes some arguments to 0 and sets the initial values for x and y. Next, the program enters a *while* loop that includes all of the math for the iterated equation. This is repeated until 128 iterations occur or until the value of $x^2 + y^2$ is greater than 4. After exiting the *while* loop, the program plots the current pixel with a color that depends on the number of iterations that occurred before the *while* loop was terminated. When the two *for* loops have finished, the program initializes arguments for the integer version and then calls *time*, which puts the current time in *second*. The program then enters two nested *for* loops and a *while* loop, which repeat the same process for the right half of the screen using integer arithmetic. When these loops have finished, the program again calls *time* to place the current time in *third*. Then *gettime* is called to place the time information in the structure *Time*, which is a structure of type *tm*. The next few statements, including the first *printf* statement, show how you can make use of the information in the structure to create a date and time statement in your own format if you'd rather do that than use one of the formats available with the functions described earlier in this chapter. The display is placed near the bottom of the screen, using the function *gotoXY* to position the cursor. (Borland C++ has a *gotoxy* function to set cursor position, but it is limited to row 24. The graphics mode that we are using has 30 rows and we are interested in using rows 27, 28, and 29, so we have to roll our own cursor positioning function. It's pretty simple, making use of the ROM BIOS video services.) Finally, the program displays the time required to generate the first half and the second half of the display. These times are obtained using the *difftime* function.

Figure 18-1. Program to Generate Phoenix Fractal

```
/*

        PHOENIX = Program to generate Phoenix Curves

                By Roger T. Stevens  8-8-92

*/

#include <dos.h>
#include <stdio.h>
#include <math.h>
#include <time.h>

union REGS reg;

void gotoXY(int col, int row);
void plot(int x,int y,int color);
void setmode(int mode);
void cls(int color);

struct dosdate_t date;
char days[7][10] = {"Sunday", "Monday", "Tuesday",
      "Wednesday","Thursday", "Friday", "Saturday"};
char months[13][10] = {"No month", "January", "February",
"March", "April", "May", "June", "July", "August",
"September", "October", "November", "December"};
char str[2] = {'A', 'P'};
int xres = 640, yres = 480, hour, i;
time_t first, second, third;
struct time Time;
int  color, row, col;
double Ymax=1.0, Ymin= -1.0, Xmax = 1.5, Xmin = -1.5;
double deltaX, deltaY, p=0.56667, q=-0.5, x, old_x, y,
```

(continued)

```
        old_y, xsq, old_xsq, ysq, xtemp, ytemp;
long int deltaP, deltaQ, pi, qi, xi, old_xi, yi, old_yi,
        xisq, yisq, xitemp, yitemp, xt, yt, oxt, oyt;
long int Pmax, Pmin, Qmax, Qmin, SCALE = 67108864L,
        MAXSIZE;

void main(void)
{
        setmode(18);
        cls(7);
        deltaX = (Xmax - Xmin)/xres;
        deltaY = (Ymax - Ymin)/yres;
        first = time(NULL);
        for (col=0; col<320; col++)
        {
                for (row=0; row<yres; row++)
                {
                        old_x = old_y = xsq = ysq = 0.0;
                        x = Ymax - row * deltaY;
                        y = Xmin + col * deltaX;
                        color = 0;
                        while ((color++<128) && ((xsq + ysq) < 4))
                        {
                                xsq = x*x;
                                ysq = y*y;
                                ytemp = 2*x*y + q*old_y;
                                xtemp = xsq - ysq + p + q*old_x;
                                old_x = x;
                                old_y = y;
                                x = xtemp;
                                y = ytemp;
                        }
                        if (color >= 128)
                                plot(col, row, 1);
                        else
                                if (color > 64)
                                        plot(col, row, 14);
                                else
                                        if(color > 48)
                                                plot(col, row, 2);
                                        else
                                                if (color > 32)
                                                        plot(col, row, 9);
                                                else
                                                        if(color > 16)
                                                                plot(col, row,
                                                                        10);
                                                        else
                                                                plot(col, row, 4);
                }
        }
        MAXSIZE = SCALE << 2;
        Pmax = 1.5 * SCALE;
```

(continued)

```
Pmin = -1.5 * SCALE;
Qmax = 1.0 * SCALE;
Qmin = -1.0 * SCALE;
pi = 0.56667 * SCALE;
qi = -0.5 * SCALE;
qi >>= 13;
deltaP = (Pmax - Pmin)/xres;
deltaQ = (Qmax - Qmin)/yres;
second = time(NULL);
for (col=320; col<xres; col++)
{
     for (row=0; row<yres; row++)
     {
          old_xi = old_yi = xisq = yisq = 0.0;
          xi = Qmax - row * deltaQ;
          yi = Pmin + col * deltaP;
          color = 0;
          while ((color++<128) && ((xisq + yisq) <
               MAXSIZE))
          {
               xt = (xi>>13);
               yt = (yi>>13);
               oxt = (old_xi>>13);
               oyt = (old_yi>>13);
               xisq = xt*xt;
               yisq = yt*yt;
               yitemp = 2*xt*yt + qi*oyt;
               xitemp = xisq - yisq + pi + qi*oxt;
               old_xi = xi;
               old_yi = yi;
               xi = xitemp;
               yi = yitemp;
          }
          if (color >= 128)
               plot(col, row, 1);
          else
               if (color > 64)
                    plot(col, row, 14);
               else
                    if(color > 48)
                         plot(col, row, 2);
                    else
                         if (color > 32)
                              plot(col, row, 9);
                         else
                              if(color > 16)
                                   plot(col, row,
                                        10);
                              else
                                   plot(col, row, 4);
     }
}
third = time(NULL);
```

(continued)

```
        getch();
        _dos_getdate(&date);
        gettime(&Time);
        hour = Time.ti_hour % 12;
        i = Time.ti_hour / 12;
        gotoXY(5,27);
        printf("Today is %s, %s %d, %d...   The time is"
             "%d:%02d" %cM.", days[date.dayofweek],
             months[date.month], date.day, date.year,
             hour,Time.ti_min, str[i]);
        gotoXY(5,28);
        printf("Floating point phoenix took %f seconds.",
             difftime(second, first));
        gotoXY(5,29);
        printf("Integer phoenix took %f seconds.",
             difftime(third, second));
        getch();
}

/*
╔══════════════════════════════════════════════════════════╗
║                                                          ║
║              setmode() = Sets video mode                 ║
║                                                          ║
╚══════════════════════════════════════════════════════════╝
*/

void setmode(int mode)
{

        union REGS reg;

        reg.x.ax = mode;
        int86 (0x10,&reg,&reg);
}

/*
╔══════════════════════════════════════════════════════════╗
║                                                          ║
║              cls() = Clears the screen                   ║
║                                                          ║
╚══════════════════════════════════════════════════════════╝
*/

void cls(int color)
{
         union REGS reg;

         reg.x.ax = 0x0600;
         reg.x.cx = 0;
         reg.x.dx = 0x1E4F;
         reg.h.bh = color;
         int86(0x10,&reg,&reg);
}
```

(continued)

```
/*
    plot() = Plots a point on the screen at a designated
        position using a selected color for 16 color
                          modes.
*/

void plot(int x, int y, int color)
{
     #define graph_out(index,val)  {outp(0x3CE,index);\
                               outp(0x3CF,val);}

     int dummy,mask;
     char far * address;

     address = (char far *) 0xA0000000L + (long)y *
     xres/8L + ((long)x / 8L);
     mask = 0x80 >> (x % 8);
     graph_out(8,mask);
     graph_out(5,2);
     dummy = *address;
     *address = color;
     graph_out(5,0);
     graph_out(8,0xFF);
}

/*
               gotoXY() = Sets cursor position
*/

void gotoXY(int col, int row)
{
     reg.h.ah = 0x02;
     reg.h.bh = 0;
     reg.h.dh = row - 1;
     reg.h.dl = col - 1;
     int86(0x10,&reg,&reg);
}
```

19

Recursion

Recursion takes place when a function calls itself. There are two types of recursion. Direct recursion occurs when *function_A* calls *function_A*. Indirect recursion occurs when *function_A* calls *function_B* (which may call other functions in a chain) until finally one of these calls *function_A* again. Recursion can be a very powerful tool in creating complex mathematical operations through very simple (but sometimes hard to understand) programming.

Whether recursion can be used with a particular language is determined by how the entire operating system and/or compiler is designed. Each time a function is called, memory is reserved to store its local variables. If a function is active and then is called again, either a whole new storage area is allocated for the local variables for the second function call, or the system assumes that the variables for this function are always stored in the same place and thus overwrites the previous set. In the first case, you can have as many copies of a function running at the same time as desired. This is the case with the C language, which permits a recurring function up to the point where system memory is exhausted. The second case is the situation with DOS. Therefore DOS cannot run functions recurrently, since the values needed for running the function in the first call are overwritten by the next call. Thus, if you are programming using DOS interrupt calls, you may not have a recurrent situation. The point, for our purposes, is that recursion is perfectly OK in C, so you may use it freely.

The von Koch Snowflake Fractal

The von Koch snowflake fractal is one of a series of fractals, called self-similar, which are produced by starting with an *initiator* consisting of some simple geometric figure composed of straight line segments, replacing each line segment with a *generator*, which is a series of line segments that make up some pattern having its beginning and end at the endpoints of the original line segment, and then repeating this process with the new line segments as many times as desired. The *generator* always has the same shape, but its size is scaled down to fit the size of the line segment being replaced for each occurrence. The von Koch snowflake begins with an *initiator* that is an equilateral triangle. The *generator* consists of four line segments that are one-third of the length of the line segment being replaced. The first and last duplicate the first and last third of the original line segment; the middle two form two sides of an equilateral triangle of which the original middle third of the line segment is the third side. Figure 19-1 is a listing of the program to create the von Koch snowflake with six line segment replacements taking place. The arrays *initiator_x1, initiator_y1, initiator_x2*, and *initiator_y2* contain the coordinates of the beginning and endpoints of the three line segments that make up the original equilateral triangle. All that the main program does is set up the graphics mode, clear the screen, and then pass each of these three line segments in turn to the function *generate*. This function begins by decrementing the parameter *level*. It then sets up the arrays *Xpoints* and *Ypoints*, which contain the coordinates of the points that must be connected to create the *generator* pattern. The first and last sets of coordinates are those of the original line segment. From these, the function determines the direction of the original line segment and sets this as the angle *theta*. It then computes the intermediate coordinates in the *generator* pattern. Next, if the parameter *level* is not 0, it recursively calls *generate* for each line segment of the pattern just computed. If *level* is 0, instead of the recursive call to *generate*, each of the coordinate sets in the coordinate arrays is plotted to the screen. (With level starting at 6, by this point in the recursion process the points are so close together that the figure appears to be con-tinuous on the screen.) The resulting fractal is shown in Figure 19-2.

As you can see, this fractal is a natural for the recursion process. Each recursion creates the coordinates for the *generator* pattern for all of the previously created line segment definitions. After the specified number of recursions has created all of the coordinates at the lowest level, they are finally plotted. This example shows clearly one very important point in using recursion. You must have some test that determines when the recursion process should terminate. Otherwise there is nothing to stop the process from repeating until you run out of memory and the system crashes. In the example, we start with a parameter called *level*, which is initialized to a value of 6. At each call to *generate*, this parameter is decremented. The recursion continues until *level* gets down to 0. At this point, the recursion process stops and the data is written to the screen. The program doesn't just terminate here, however; all that happens is that this particular copy of the function *generate* is completed, causing the program to return to the calling point and continue until finally all of the copies of *generate* have been completed, at which point the program finally terminates.

Figure 19-1. Listing of Program to Generate von Koch Snowflake

```
/*

        SNOWFLAK = PROGRAM TO GENERATE KOCH SNOWFLAKE

                  By Roger T. Stevens   6-17-92

*/

#include <stdio.h>
#include <math.h>
#include <dos.h>

void cls(int color);
void generate (float X1, float Y1, float X2, float Y2, int
      level);
void plot(int x, int y, int color);
void setmode(int mode);

int i;
int generator_size = 5;
int xres=640, yres=480;
int level=6;
int initiator_x1[10] =
{170,320,470},initiator_x2[10]={320,
      470,170},initiator_y1[10]={165,425,165},
      initiator_y2[10]={425,165,165};
```

(continued)

```
float r, theta;
main()
{
      setmode(18);
      cls(0);
      for (i=0; i<3; i++)
      {
            generate(initiator_x1[i], initiator_y1[i],
                  initiator_x2[i], initiator_y2[i],
                  level);
      }
      getch();
}

/*
╔══════════════════════════════════════════════════════════════╗
║                                                              ║
║              generate() = Generates curve                    ║
║                                                              ║
╚══════════════════════════════════════════════════════════════╝
*/

void generate (float X1, float Y1, float X2, float Y2, int
      level)
{
      int j, k;
      float Xpoints[5], Ypoints[5];

      level--;
      r = (sqrt((X2 - X1)*(X2 - X1) + (Y2 - Y1)*(Y2 -
            Y1)))/3.0;
      Xpoints[0] = X1;
      Ypoints[0] = Y1;
      Xpoints[4] = X2;
      Ypoints[4] = Y2;
      if ((X2 - X1) == 0)
            if (Y2 > Y1)
                  theta = 90;
            else
                  theta = 270;
      else
            theta = atan((Y2-Y1)/(X2-X1))*57.295779;
      if (X1>X2)
            theta += 180;
      Xpoints[1] = X1 + r*cos(theta*.017453292);
      Ypoints[1] = Y1 + r*sin(theta*.017453292);
      theta += 60;
      Xpoints[2] = Xpoints[1] + r*cos(theta*.017453292);
      Ypoints[2] = Ypoints[1] + r*sin(theta*.017453292);
      theta -= 120;
      Xpoints[3] = Xpoints[2] + r*cos(theta*.017453292);
      Ypoints[3] = Ypoints[2] + r*sin(theta*.017453292);
      if (level > 0)
      {
```

(continued)

```
            for (j=0; j<generator_size-1; j++)
            {
                  X1 = Xpoints[j];
                  X2 = Xpoints[j+1];
                  Y1 = Ypoints[j];
                  Y2 = Ypoints[j+1];
                  generate (X1,Y1,X2,Y2,level);
            }
      }
      else
      {
            for (k=0; k<generator_size; k++)
            {
                  plot(Xpoints[k],Ypoints[k],15);
            }
      }
}

/*
                 setmode() = Sets video mode
*/

void setmode(int mode)
{
      union REGS reg;

      reg.x.ax = mode;
      int86 (0x10,&reg,&reg);
}

/*
                 cls() = Clears the screen
*/

void cls(int color)
{
       union REGS reg;

       reg.x.ax = 0x0600;
       reg.x.cx = 0;
       reg.x.dx = 0x1E4F;
       reg.h.bh = color;
       int86(0x10,&reg,&reg);
}
```

(continued)

```
/*
 ┌──────────────────────────────────────────────────────────┐
 │                                                          │
 │  plot() = Plots a point on the screen at a designated    │
 │    location in a selected color for 16 color modes.      │
 │                                                          │
 └──────────────────────────────────────────────────────────┘
*/

void plot(int x, int y, int color)
{
      #define graph_out(index,val)  {outp(0x3CE,index);\
                                     outp(0x3CF,val);}

      int dummy,mask;
      char far * address;

      address = (char far *) 0xA0000000L + (long)y *
            xres/8L + ((long)x / 8L);
      mask = 0x80 >> (x % 8);
      graph_out(8,mask);
      graph_out(5,2);
      dummy = *address;
      *address = color;
      graph_out(5,0);
      graph_out(8,0xFF);
}
```

Figure 19-2. von Koch Snowflake Fractal

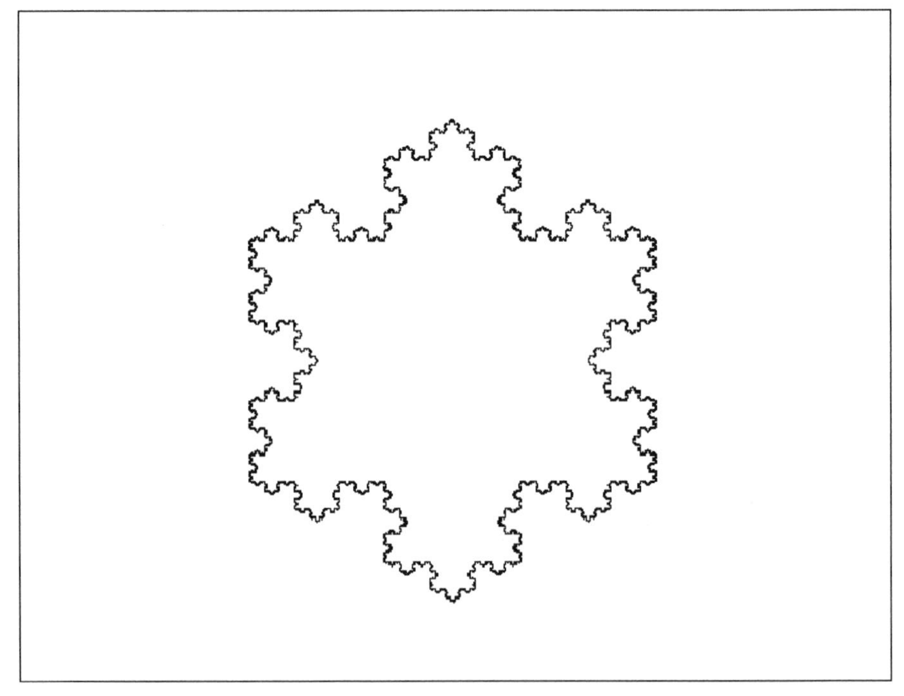

20

How to Handle Frequently Used Functions

If you've been working closely with the programs described in the previous chapters of this book, you're probably getting pretty tired of some of the functions like *setmode, cls* and *plot* that seem to be repeated in just about every program. You're probably thinking that there must be a better way to work with these functions than to have rewrite them in every new program that you're designing to create a new fractal. The reason these functions are repeated for each program in this book is so that you can have programs that stand alone, and can be run without referring to anything external. In this way, you can concentrate on the program and lesson at hand and not have to go leafing back through the book, trying to find functions from previous sections that are a necessary part of your present program. However, you're right in thinking that there are ways to avoid continual repetition of functions that are used frequently in a lot of different programs. This chapter will describe how this is done.

The Seventh-Order Newton's Method Fractal

As an example of the available techniques, we're going to use a Newton's method fractal program. How Newton's method is used to create fractals is described in detail in Chapter 15, using the third-order fractal as an example. The fractal generated here is the one that is obtained in trying to obtain the roots of the equation

$$z^7 - 1 = 0 \qquad \text{(Equation 20-1)}$$

by guessing at a root value and then applying the iterated equation

$$\begin{aligned} z_n &= z_{n-1} - \frac{f(z_{n-1})}{f'(z_{n-1})} \\ &= \frac{6z_{n-1}}{7} + \frac{1}{7z_{n-1}^6} \end{aligned} \qquad \text{(Equation 20-2)}$$

The program is listed below in Figure 20-1 and the resulting fractal picture is shown in Plate 16.

Figure 20-1. Seventh-Order Newton's Method Fractal Generating Program

```
/*

    NEWTON7 = Map of Newton's method for solving z7 = 1

            By Roger T. Stevens   8-23-92

*/

#include <dos.h>
#include <stdio.h>
#include <math.h>
#include <stdlib.h>
#include "graphics.c"

int xres = 640, yres = 480;
int color, i, row, col;
char number[15], *end=number;
float Pmax=1.4, Pmin= -1.4, Qmax = 1.05, Qmin = -1.05;
double P, deltaP, deltaQ, denom, x, y, old_x, old_y, xsq,
      ysq, ytemp, sqsum;

void main(void)
{
      setmode(18);
      cls(7);
      deltaP = (Pmax - Pmin)/xres;
```

(continued)

```
        deltaQ = (Qmax - Qmin)/yres;
        for(col=0; col<xres; col++)
        {
                for(row=0; row<yres; row++)
                {
                        x = Pmin + col*deltaP;
                        y = Qmax - row*deltaQ;
                        xsq = ysq = 0;
                        old_x = old_y = 42;
                        for (i=0; i<64; i++)
                        {
                                xsq = x*x;
                                ysq = y*y;
                                sqsum = xsq + ysq;
                                denom = 7.0*(sqsum*sqsum*sqsum
                                        *sqsum*sqsum*sqsum);
                                if ((denom > -.00004) && (denom <
                                        .00004))
                                        denom = .00004;
                                ytemp = 0.85714285*y - (6.0*x*x*x
                                        *x*x*y - 20.0*x*x*x*y*y*y +
                                        6.0*x*y*y*y*y*y)/denom;
                                x = 0.85714285*x + (x*x*x*x*x*x -
                                        15.0*x*x*x*x*y*y +
                                        15.0*x*x*y*y*y*y -
                                        y*y*y*y*y*y)/denom;
                                y = ytemp;
                                if (fabs(old_x - x) < 1E-10 &&
                                        (fabs(old_y - y) < 1E-10))
                                        break;
                                old_x = x;
                                old_y = y;
                        }
                        plot(col, row, i%16);
                }
        }
        getch();
}
```

You'll observe that the structure of the program is very similar to the Newton's method program for the third-order fractal, except for the necessary changes in mathematics. However, you can heave a sigh of relief, for the old standard *setmode, cls*, and *plot* functions are gone. How we handle them in several different ways will be described in the rest of this chapter.

Functions in an *include* Statement

Old time C programmers are convinced that nothing should ever be defined by an *include* statement except a header file. Thus the expression

```
#include "graphics.c"
```

in the program above will be anathema to them. Nonetheless, it works very well. We have a *graphics.c* program that is listed in Figure 20-2. It includes all of those frequently used functions. When we compile the *newton7.c* program, the compiler includes and compiles the entire *graphics.c* program also, providing those frequently used functions that we need to make the program operative. This has the distinct advantage that you don't need to modify you *newton7.c* program in any way except to have the *include* statement in it. It has the disadvantage that the compiler has to compile the *graphics.c* program again every time the *newton7.c* program is recompiled, even if the *graphics.c* program is never altered. If you have a program that uses a lot of functions that never require changes, you can save a lot of compilation time by using another technique than this one.

Now, a word about where you are going to put the *graphics.c* program. In C, the quotation marks around the program name in the *include* statement means that the compiler looks for this program in the same directory where you have your main program (that program is *newton7.c* in this case). If you have the program somewhere else, the compiler won't be able to find it. There is another alternative. You can use a pair of carets <> instead of the quotations marks, as was done for the *include* statements for regular C library header files. In this case, the compiler looks for your program in the same directory with all of the header files. For Borland C++, this directory is usually *c:\borlandc\include*. Your C graphics program may be a little lonely in there with all of the header files, but the compiler will find it without difficulty.

Figure 20-2. Listing of the graphics.c Program

```
/*
 ╔════════════════════════════════════════════════════════════╗
 ║                                                            ║
 ║  graphics.c = Funtions to perform graphics operations      ║
 ║                                                            ║
 ║             By Roger T. Stevens  8-23-92                   ║
 ║                                                            ║
 ╚════════════════════════════════════════════════════════════╝
*/

#include <dos.h>
```

(continued)

```
extern xres;

/*
        setmode() = Sets video mode
*/

void setmode(int mode)
{
      union REGS reg;

      reg.x.ax = mode;
      int86 (0x10,&reg,&reg);
}

/*
        cls() = Clears the screen
*/

void cls(int color)
{
       union REGS reg;

       reg.x.ax = 0x0600;
       reg.x.cx = 0;
       reg.x.dx = 0x1E4F;
       reg.h.bh = color;
       int86(0x10,&reg,&reg);
}

/*
   plot() = Plots a point on the screen at a designated
       position using a selected color for 16 color
                         modes
*/

void plot(int x, int y, int color)
{
      #define graph_out(index,val)  {outp(0x3CE,index);\
                                     outp(0x3CF,val);}

      int dummy,mask;
      char far * address;
```

(continued)

```
        address = (char far *) 0xA0000000L + (long)y *
              xres/8L + ((long)x / 8L);
        mask = 0x80 >> (x % 8);
        graph_out(8,mask);
        graph_out(5,2);
        dummy = *address;
        *address = color;
        graph_out(5,0);
        graph_out(8,0xFF);
}
```

Creating a Library

Another way of handling your frequently used functions is to put them in a library. In this case, you compile the *graphics.c* file into a *graphics.obj* file. You also need a header file that defines the functions that are in the library. Figure 20-3 shows a listing for the header file for our graphics library. This header file is named in the *include* statement, instead of the *.c* file we previously used. The statement becomes

```
#include "graphics.h"
```

Figure 20-3. Header File for Graphics Library

```
/*

     graphics.h = Header file for graphics operations
               By Roger T. Stevens  8-23-92

*/

void setmode(int mode);
void cls(int color);
void plot(int x, int y, int color);
```

The big advantage of this technique is that the compiler can use the library functions in any program without having to recompile them. It just links the object code directly from the graphics library. A second advantage is that the compiler will only get the object code from the library for those functions that are actually referenced by the using program. Functions not used are omitted, reducing the size of the program. There are a couple of disadvantages, however. First, if you make a library with functions that have been compiled using one compiler, it usually will not work when linked to a program that is compiled with a different compiler. This is often true when the new compiler is the next upgrade to the one that you have been using, so that most likely, if you get a new compiler or an upgrade to your old one, you'll have to recompile your library of functions and then rebuild the library. Second, while programs that run in an integrated environment (such as those used by Borland C++) can run just fine using the C libraries, they cannot run in the same way with your home-built libraries. Instead of just running the program, you have to set up a project file that contains the path and name of your library and of your main program. Then, when you run the project file, the compiler will be able to locate your library and link the functions properly. (Unfortunately, when you get one project file in your directory, it has a bad habit of running when you want to run some other program.)

Once you have compiled all the programs that you want in the library, and have *.OBJ* files for them, you are ready to create the library. The program (in Borland C++) that is used to generate the library is *TLIB*. The syntax for running it is

```
TLIB libname [/C] [/E] [/P] [/O] commands, listfile
```

The parameter *libname* is the name that you plan to assign to your library file. It must be an acceptable DOS file name. The four options in brackets have the following meaning:

`/C`	make the library case sensitive
`/E`	create an extended dictionary
`/PSIZE`	set the library page size to SIZE
`/O`	purge all comment records

The commands are a symbol (which tells what to do with an object module) followed by the name of the object module. The acceptable commands are:

`+`	add object module to the library
`-`	remove the object module from the library
`*`	extract module without removing it
`-+` or `+-`	replace object module in library

`-*` or `*-` extract object and remove it

In addition, you may use *@filename* to continue with commands, etc. from the file *filename*. You may also use & at the end of a line to continue additional commands and options on the next line. The line that was used to generate a *graphics.lib* library was

```
TLIB graphics.lib + graphics.obj
```

21

Common Mistakes in C Programming

C is one of the most versatile and flexible of computer languages. Sometimes, however, the freedom that C gives you as a programmer can result in your making subtle mistakes that cause your program to behave in strange and unusual ways. It is not always easy to look at a program listing and detect such errors. Sometimes you can spend hours searching for the source of some problem only to discover some simple silly mistake. This chapter gives illustrations of a number of mistakes that are pretty common. Even experienced programmers make these errors at times and often spend a lot of time hunting for them. If you've having a problem with unusual behavior of your program, refer to this chapter and you may find that you have made one of these simple errors.

Using the Wrong Units

Suppose that you have written this program:

```
#include <stdio.h>
#include <conio.h>
#include <math.h>

void main(void)
{
	double cosine;

	cosine = cos(45.0);
	printf("\ncosine of 45 degrees is"
		" %2.10lf",cosine);
	getch();
}
```

What you expected to see printed out was the cosine of 45 degrees. Unfortunately, the *cos* mathematical function requires an argument expressed in radians. Consequently, instead of displaying the cosine of 45 degrees, you are displaying the argument of 45 radians, which is something totally different from what you want.

Failure to Include Proper Header Files

The following program corrects the error in the previous one by converting the degrees to radians. However, it has an error of its own. In this program, you failed to include the *math.h* header at the beginning of your program. During compilation and running of the program, you get no indication that anything is wrong, but the value returned by the *cos* function is pure garbage if the statement *#include <math.h>* is missing. It is important to include the proper header statement for every function that you use from the C libraries. If you leave out a needed header statement, the reaction of the program and compiler differs, depending on the circumstances, but the end result is always disastrous.

```
#include <stdio.h>
#include <conio.h>

void main(void)
{
      double cosine;

      cosine = cos(45.0 * 0.0174533125);
      printf("\ncosine of 45 degrees is"
           " %2.10lf",cosine);
      getch();
}
```

Wrong Expression for Equality in Conditional Statements

In the program below, you expected to display the value of *i* for values less than or equal to 10. For values over 10, you wanted to display *i is larger than 10*. What actually happens is that you print out *i is larger than 10* once and that is all. Why is this? Your *if* statement should have been written *if (i == 11)* to do what you wanted. When you wrote it with a single equals sign, instead of testing for equality, it sets the value of *i* to 11. The conditional statement is true, so the legend *i is larger than 10* is displayed and then since *i* is now 11, the loop terminates at the beginning of the next pass. Similar peculiar results will occur whenever you make the mistake of using = when you should be using == in an *if* statement. Sometimes you'll get a compiler warning about this; sometimes you won't.

```
#include <stdio.h>
#include <conio.h>

void main(void)
{
      int i, no;

      for (i=0; i<12; i++)
      {
            if (i = 11)
                  printf("\ni is larger than 10");
            else
                  printf("\ni is %d",i);
      }
      getch();
}
```

Improper Use of Semicolon

Look at the following program

```
#include <stdio.h>
#include <conio.h>

void main(void)
{
      int i, no;

      for (i=0; i<12; i++);
      {
                  printf("\ni is %d",i);
      }
      getch();
}
```

You wanted to display each value of *i* from 0 to 11. Instead all you got was a single line that said *i is 12*. The reason is that you put a semicolon after the *for* statement. If you put a semicolon after an *if*, *for*, or *while* statement, this terminates the statement. Thus the *for* loop in this case iterates for *i=0* to *i=11* without doing anything. When *i=12* the program exits the loop and performs the *printf* statement, which because of the misplaced semicolon is not part of the *for* loop.

Failure to Allow Loop to Terminate

In the following program you wanted to display values of *i* from 0 through 9. Instead, the program keeps repeating the value for *i=0* over and over until you stop it. The reason is that you forgot to increment *i* somewhere within the *while* loop. When you use a *for* loop, the contruction of the *for* statement takes care of the increasing of the index variable, but when you use a *while* or *do-while* loop, it is your responsibility to provide for this increasing of the index. You could have done this by changing the statement within the *while* loop to *printf("\ni = %d",i++);*.

```
#include <stdio.h>
#include <conio.h>

void main(void)
{
        int i=0;

        while (i<10)
        {
                printf("\ni = %d",i);
        }
        getch();
}
```

Wrong Format for *printf* Function

In the following program, you attempt to display a floating point number using the *%d* (integer) format. The result is that each time you display a 0 instead of the actual value.

```
#include <stdio.h>
#include <conio.h>

void main(void)
{
        int i=0;
        float k;

        while (i<10)
        {
                k = i++;
                printf("\nk = %d",k);
        }
        getch();
}
```

Improper Definition of Far Pointer

In the program below, you attempt to define two pointers to far arrays of integers. Unfortunately, the *far* only applies to the first argument, not the second. Although the program sets aside the proper memory area for each buffer with *farmalloc*, only the first pointer can properly contain the far address. Thus when you run the program, it will stuff data into the wrong memory area, overlapping some important parts of the program that are already there. The result is that the program will go off to never-never land instead of initializing the array and displaying a value as you had intended.

```
#include <stdio.h>
#include <conio.h>
#include <alloc.h>

void main(void)
{
      unsigned int far *buf1, *buf2;
      unsigned int i;

      buf1 = farmalloc(66000);
      if (buf1 == NULL)
      {
            printf("\nNot enough memory for"
                  " buf1...");
            getch();
            exit(0);
      }
      buf2 = farmalloc(66000);
      if (buf2 == NULL)
      {
            printf("\nNot enough memory for"
                  " buf2...");
            getch();
            exit(0);
      }
      for (i=0; i<33000; i++)
      {
            buf2[i] = i;
      }
      printf("\nbuf2[31234] = %d",buf2[31234]);
      getch();
}
```

Exceeding Array Limits

In the following program, the array *j* only has 8 members, yet you are trying to display 10 values. C doesn't mind this at all; it displays the values of the 8 legitimate members of the array and the integer values that are in the next memory locations after the end of the array. This doesn't do much harm here; you just get a couple of garbage values in your display. If you were writing to these array members instead of reading from them, you'd put something unexpected into memory locations that were reserved for other arguments, which might result in all kinds of bizarre behavior by the program.

```
#include <stdio.h>
#include <conio.h>

void main(void)
{
        int i,j[8]={1,2,3,4,5,6,7,8};

        for (i=0; i<10; i++)
        {
                printf("\nj[%d] = %d",i,j[i]);
        }
        getch();
}
```

Problems with *#define* Statements

Near the beginning of your program, you have the following *#define* statement:

```
#define graph_out(index,val)  {outp(0x3CE,index);\
                              outp(0x3CF,val);}
```

Later on in the program, this occurs:

```
if (a == 4)
        graph_out(index,val);
else
        a++;
```

When you compile the program, you are given an error message which tells you that you have a misplaced *else*. Why? When the compiler encounters the *graph_out* statement following the *if*, it replaces it by the contents of the *#define* statement at the beginning of the program.

This consists of two lines of code. Only the first line is considered to be a part of the preceeding *if* statement. The second line is just an ordinary statement that is run regardless of whether the *if* is true or false. Thus when the compiler encounters the *else* it doesn't accept it as an alternative to be performed instead of the *if* so it gives the error message. What you need to do, is surround the *graph_out* statement with curly brackets, since to the compiler it is more than one line, even though it doesn't appear to be so when you look at the listing. Then everything will work out all right. You also need to be thankful that you used the *if-else* construction. If you had just used the *if*, only the first line of the *#define* statement would have been associated with the *if*, but the code would have compiled without an error message and you might have spent a lot of time trying to track down why your program wasn't doing quite what you wanted it to do.

ASCII Output Characters

Decimal	Hexa-decimal	Char-acter	Decimal	Hexa-decimal	Char-acter
00	00		16	10	▸
01	01	☺	17	11	◂
02	02	☻	18	12	↕
03	03	♥	19	13	‼
04	04	♦	20	14	¶
05	05	♣	21	15	§
06	06	♠	22	16	▬
07	07	•	23	17	↨
08	08	◘	24	18	↑
09	09	○	25	19	↓
10	0A	◙	26	1A	→
11	0B	♂	27	1B	←
12	0C	♀	28	1C	∟
13	0D	♪	29	1D	↔
14	0E	♫	30	1E	▲
15	0F	☼	31	1F	▼

Decimal	Hexa-decimal	Char-acter	Decimal	Hexa-decimal	Char-acter
32	20		56	38	8
33	21	!	57	39	9
34	22	"	58	3A	:
35	23	#	59	3B	;
36	24	$	60	3C	<
37	25	%	61	3D	=
38	26	&	62	3E	>
39	27	'	63	3F	?
40	28	(	64	40	@
41	29	)	65	41	A
42	2A	*	66	42	B
43	2B	+	67	43	C
44	2C	,	68	44	D
45	2D	-	69	45	E
46	2E	.	70	46	F
47	2F	/	71	47	G
48	30	0	72	48	H
49	31	1	73	49	I
50	32	2	74	4A	J
51	33	3	75	4B	K
52	34	4	76	4C	L
53	35	5	77	4D	M
54	36	6	78	4E	N
55	37	7	79	4F	O

<table>
<tr><th>Decimal</th><th>Hexa-decimal</th><th>Char-acter</th><th>Decimal</th><th>Hexa-decimal</th><th>Char-acter</th></tr>
<tr><td>80</td><td>50</td><td>P</td><td>104</td><td>68</td><td>h</td></tr>
<tr><td>81</td><td>51</td><td>Q</td><td>105</td><td>69</td><td>i</td></tr>
<tr><td>82</td><td>52</td><td>R</td><td>106</td><td>6A</td><td>j</td></tr>
<tr><td>83</td><td>53</td><td>S</td><td>107</td><td>6B</td><td>k</td></tr>
<tr><td>84</td><td>54</td><td>T</td><td>108</td><td>6C</td><td>l</td></tr>
<tr><td>85</td><td>55</td><td>U</td><td>109</td><td>6D</td><td>m</td></tr>
<tr><td>86</td><td>56</td><td>V</td><td>110</td><td>6E</td><td>n</td></tr>
<tr><td>87</td><td>57</td><td>W</td><td>111</td><td>6F</td><td>o</td></tr>
<tr><td>88</td><td>58</td><td>X</td><td>112</td><td>70</td><td>p</td></tr>
<tr><td>89</td><td>59</td><td>Y</td><td>113</td><td>71</td><td>q</td></tr>
<tr><td>90</td><td>5A</td><td>Z</td><td>114</td><td>72</td><td>r</td></tr>
<tr><td>91</td><td>5B</td><td>[</td><td>115</td><td>73</td><td>s</td></tr>
<tr><td>92</td><td>5C</td><td>\</td><td>116</td><td>74</td><td>t</td></tr>
<tr><td>93</td><td>5D</td><td>]</td><td>117</td><td>75</td><td>u</td></tr>
<tr><td>94</td><td>5E</td><td>^</td><td>118</td><td>76</td><td>v</td></tr>
<tr><td>95</td><td>5F</td><td>_</td><td>119</td><td>77</td><td>w</td></tr>
<tr><td>96</td><td>60</td><td>‘</td><td>120</td><td>78</td><td>x</td></tr>
<tr><td>97</td><td>61</td><td>a</td><td>121</td><td>79</td><td>y</td></tr>
<tr><td>98</td><td>62</td><td>b</td><td>122</td><td>7A</td><td>z</td></tr>
<tr><td>99</td><td>63</td><td>c</td><td>123</td><td>7B</td><td>{</td></tr>
<tr><td>100</td><td>64</td><td>d</td><td>124</td><td>7C</td><td>|</td></tr>
<tr><td>101</td><td>65</td><td>e</td><td>125</td><td>7D</td><td>}</td></tr>
<tr><td>102</td><td>66</td><td>f</td><td>126</td><td>7E</td><td>~</td></tr>
<tr><td>103</td><td>67</td><td>g</td><td>127</td><td>7F</td><td>⌂</td></tr>
</table>

Decimal	Hexa-decimal	Char-acter	Decimal	Hexa-decimal	Char-acter
128	80	Ç	152	98	ĳ
129	81	ü	153	99	Ö
130	82	é	154	9A	Ü
131	83	â	155	9B	¢
132	84	ä	156	9C	£
133	85	à	157	9D	¥
134	86	å	158	9E	₧
135	87	ç	159	9F	ƒ
136	88	ê	160	A0	á
137	89	ë	161	A1	í
138	8A	è	162	A2	ó
139	8B	ï	163	A3	ú
140	8C	î	164	A4	ñ
141	8D	ì	165	A5	Ñ
142	8E	Ä	166	A6	ª
143	8F	Å	167	A7	º
144	90	É	168	A8	¿
145	91	æ	169	A9	⌐
146	92	Æ	170	AA	¬
147	93	ô	171	AB	½
148	94	ö	172	AC	¼
149	95	ò	173	AD	¡
150	96	û	174	AE	«
151	97	ù	175	AF	»

Decimal	Hexa-decimal	Char-acter	Decimal	Hexa-decimal	Char-acter
176	B0	░	200	C8	╚
177	B1	▒	201	C9	╔
178	B2	▓	202	CA	╩
179	B3	│	203	CB	╦
180	B4	┤	204	CC	╠
181	B5	╡	205	CD	═
182	B6	╢	206	CE	╬
183	B7	╖	207	CF	╧
184	B8	╕	208	D0	╨
185	B9	╣	209	D1	╤
186	BA	║	210	D2	╥
187	BB	╗	211	D3	╙
188	BC	╝	212	D4	╘
189	BD	╜	213	D5	╒
190	BE	╛	214	D6	╓
191	BF	┐	215	D7	╫
192	C0	└	216	D8	╪
193	C1	┴	217	D9	┘
194	C2	┬	218	DA	┌
195	C3	├	219	DB	█
196	C4	─	220	DC	▄
197	C5	┼	221	DD	▌
198	C6	╞	222	DE	▐
199	C7	╟	223	DF	▀

Decimal	Hexa-decimal	Char-acter	Decimal	Hexa-decimal	Char-acter
224	E0	α	240	F0	≡
225	E1	ß	241	F1	±
226	E2	Γ	242	F2	≥
227	E3	π	243	F3	≤
228	E4	Σ	244	F4	⌠
229	E5	σ	245	F5	⌡
230	E6	µ	246	F6	÷
231	E7	τ	247	F7	≈
232	E8	Φ	248	F8	°
233	E9	Θ	249	F9	·
234	EA	Ω	250	FA	·
235	EB	δ	251	FB	√
236	EC	∞	252	FC	ⁿ
237	ED	φ	253	FD	²
238	EE	ε	254	FE	■
239	EF	∩	255	FF	

Index

Other Academic Press Titles of Interest

The Complete C++ Primer, Second Edition
by Keith Weiskamp and Bryan Flamig

This new edition of the popular hands-on guide to C++ programming provides an easy way to help you master C++. Let noted authors Keith Weiskamp and Bryan Flamig show you how to use the important features of C++ Version 2.1, including classes, functions, constructors and destructors, stream I/O, operator overloading, inheritance, and more. You'll also find coverage of the new template features.

ISBN 0-12-742686-8 $34.95

The C Graphics Handbook
by Roger T. Stevens

Programming graphics in C is made easy with *The C Graphics Handbook*. This handbook contains all of the tools needed to set up display modes for EGA, VGA, or Super VGA cards. It also covers three-dimensional drawing techniques using C and C++, and provides programs for saving display screens to disk files and restoring screens using the common compression formats PCX, IMG, and GIF. Also includes a disk for IBM and compatibles containing all of the code from the text.

ISBN 0-12-668320-4 $39.95

Fractals Everywhere
by Michael Barnsley

This text focusses on how fractal geometry can be used to model real objects in the physical world. Rather than considering only fractal images that have been generated randomly, the approach here is to start with a natural object and find a specific fractal to fit it. Applications of fractal geometry extend to biological modeling, physiological modeling, geography, coastlines, turbulence, images, computer graphics, feathers, and ocean spray. The applications to computer graphics and, in particular, image compression for data transmission and reconstruction, are exciting new developments.

ISBN 0-12-079062-9 $49.95

Other Academic Press Titles of Interest

The Desktop Fractal Design System
by Michael Barnsley

Zoom-in and explore the world of fractal design with *The Desktop Fractal Design System*. Designed to complement Barnsley's *Fractals Everywhere*, this system will be an indispensable educational and scientific tool for students, engineers, and scientists in all disciplines. The software helps to connect theoretical concepts with on-screen geometric modeling.

ISBN (Macintosh)	0-12-079065-3	$49.95
ISBN (IBM)	0-12-079063-7	$49.95

Fractal Attraction
by Kevin D. Lee and Yosef Cohen

Design an unlimited number of fractal images using iterated function systems (IFS); math degree not required! In a draw-like window, use the power of Macintosh to build an IFS code just by pointing, clicking, and dragging. *Fractal Attraction* automatically calculates and displays the IFS equations in a spreadsheet-type window. The resulting image is generated in the fractal window.

ISBN 0-12-440740-4 $49.95

ORDER FORM

To Order: Return this form with your payment to Academic Press, Order Fulfillment Department, 6277 Sea Harbor Drive, Orlando, FL 32821-9816, or **call toll-free 1-800-321-5068.**

QUANTITY	AUTHOR/TITLE	ISBN	PRICE
		Subtotal	
		Sales Tax (where applicable)	
		TOTAL	

☐ Payment enclosed (please include applicable tax)

☐ Bill me directly (We cannot ship to a P.O. box)*

☐ Bill my company (purchase order attached)*

*Shipping, handling, and tax will be added to billed orders. Tax will be added to credit card orders.

Charge card #____________________ Expiration Date __________

Your Signature______________________________

Name____________________ Telephone__________

Address ______________________________

City____________________ State/Country__________

Zip/Postal Code____________________